# COGENERATION SOURCEBOOK

# COGENERATION SOURCEBOOK

Compiled and Edited by
F. William Payne

Published by
**THE FAIRMONT PRESS, INC.**
P.O. Box 14227
Atlanta, Georgia 30324

*COGENERATION SOURCEBOOK*

Published by The Fairmont Press, Inc., P.O. Box 14227, Atlanta, Georgia 30324.


Library of Congress Catalog Card No. 84-48530
ISBN: 0-88173-002-5

**Library of Congress Cataloging in Publication Data**
Main entry under title:

Cogeneration sourcebook.

   Includes bibliographies and index.
   1. Cogeneration of electric power and heat. I. Payne,
F. William, 1924-
TK1041.C634   1985    333.79'3   84-48530
ISBN 0-88173-002-5

DISTRIBUTED:   Europe, Japan, India, Middle East, Southeast Asia and Africa by E. & F.N. Spon, 11 New Fetter Lane, London EC4PEE.

Latin America, Australia, New Zealand, China and Iron Curtain Countries by Feffer & Simons, Inc., 100 Park Avenue, New York, NY 10017.

Canada and U.S. (Libraries and Bookstores) by Van Nostrand Reinhold, 135 W. 50th Street, New York, NY 10020.

# CONTENTS

# CONTRIBUTORS

R.T. Baltes, P.E.
Baltes/Valentino Associates, Ltd.

R.N. Basu
University of Calgary

Fred E. Becker
Thermo Electron

Wallace E. Brand, J.D.
Brand & Leckie

Kenneth M. Clark, P.E.
Burns & McDonnell Engineering
    Company

James R. Clements
United Enertec, Inc.

L.L. Cogger
University of Calgary

Douglas E. Criner
Burns & McDonnell Engineering
    Company

Craig R. Cummings
Science Applications, Inc.

Gerald Decker, Ph.D.
Decker Energy International, Inc.

Joseph P. Fellner, Lt.
Wright-Patterson AFB

Roger L. Froneberger
Oklahoma State University

Vijay K. Gupta, Ph.D.
Central State University

Wayne E. Hanson
CH2M Hill

Dallas Jones
Illinois Department of Energy

Azmi Kaya
Bailey Controls Company

M.A. Keyes
Bailey Controls Company

Azam Khan, P.E.
Illinois Department of Energy

Fred J. Krause
Georgia Power Company

Richard F. Kurzynske, Ph.D.
Gas Research Institute

R.H. McMahan, Jr.
General Electric Company

Martin A. Mozzo, Jr., P.E.
American Standard, Inc.

William J. Murphy, P.E.
Baltes/Valentino Associates, Ltd.

E.J. Ney
Georgia Power Company

Steven A. Parker
Oklahoma State University

W.G. Reed
General Electric Company

Ravi Sakhuja, Ph.D.
Thermo Electron

Alan C. Sommer, P.E.
Bailey Controls Company

Robin W. Taylor, P.E.
Science Applications, Inc.

Clayton R. Thompson
Oklahoma State University

Eric Thompson
Washington University

Steven L. Tuma, P.E., C.E.M.
Illinois State Buildings Energy
    Program

Wayne C. Turner, Ph.D., P.E.,
    C.E.M.
Oklahoma University

MacAuley Whiting, Jr.
Decker Energy International, Inc.

Claire Wooster
Washington University

# FOREWORD

The *Cogeneration Sourcebook* includes the latest information on cogeneration planning, financing, and technical improvements. Each chapter is timely, topical, up-to-the minute; the authors are leading practitioners in the burgeoning cogeneration industry.

Several vital new approaches to cogeneration are covered, including the growth of prepackaged and small-scale systems. Developmental concepts such as solar cogeneration systems, fuel cell cogeneration systems, and other renewable energy cogeneration systems are reviewed by energy professionals directly responsible for the programs.

New techniques of financing cogeneration systems are discussed, as are the latest regulatory procedures required for successful implementation.

The *Cogeneration Sourcebook* is an essential reference for all energy specialists and managers who must keep up to date on the changes taking place in the multi-billion dollar cogeneration industry.

F. William Payne, Editor-in-Chief
*Strategic Planning and Energy Management*

# CHAPTER 1
# Cogeneration — An Energy Conservation and Cost Savings Update

*V. Gupta*

## COGENERATION AND LEGISLATION

The National Energy Act, a five-piece legislative package, was signed into law by President Carter in November, 1978. This legislative package represented a strong effort on the part of the Federal Government to lay a solid foundation for a comprehensive national energy policy. Each part of NEA has a Public Law number and a title, and they are as follows:

Public Utility Regulatory Policies Act - P.L.95-617

Energy Tax Act of 1978 - P.L.95-618

National Energy Conservation Policy Act - P.L.95-619

Powerplant and Industrial Fuel Use Act - P.L.95-620

Natural Gas Policy Act - P.L.95-621.

The provisions of the NEA are expected to result in reduced oil import needs, increased use of fuels other than oil and gas, and more efficient and more equitable use of energy in the United States.

The Public Utilities Regulatory Policy Act (PURPA) in particular provides significant incentives for cogeneration technology. The main features of PURPA in relation to congeneration are as follows:[1]

- Qualified cogenerators are exempted from huge state and federal regulations that are applicable to utilities.

- Qualified cogenerators have a right to a connection to the grid of an electric utility company.

1

- Electric utilities must provide standby or back up electric power to the cogenerators under non-discriminatory rates and policies.

- Electric utilities are required to buy or sell power from qualified cogenerators at just and reasonable rates.

- Industries are in little peril of being publicly labeled as utilities.

The above policies present an altogether different viewpoint for the advancement of cogeneration technology as compared to the effort of 1960's. Many other state and federal initiatives also provide various incentives for cogeneration. The New York State Cogeneration Act of 1980 states: "It is in the public interest to encourage the development of cogeneration facilities in order to conserve our finite and expensive energy resources and to provide for their most efficient utilization." This important legislation which may set an example for other states, exempts "cogeneration facilities" from state and local permits, various construction requirements, and operational conditions.

In essence, the New York legislation is designed to promote cogeneration in industries and involve utilities. This state is one among many nationwide taking cognizance of what can be done and what can be gained with this off-the-shelf technology.

"Slowly, the regulatory system is adapting to the needs of conservation in general, and electricity rates are in the process of being revised so that they encourage, rather than discourage, cogeneration," wrote Yergin. "Altogether, it may be economically possible to cut industrial energy use by more than a third through cogeneration and conservation efforts."[2] As much as $40-billion in total capital investment could be saved by industry with emphasis on cogeneration and conservation compared with the capital investments necessary with conventional energy conversion approaches.

## THERMODYNAMIC CONSIDERATIONS OF COGENERATION

The second law of thermodynamics tells us that quality of energy can change only in one direction and that energy loses its capacity to do useful work, ultimately reaching the point of zero usefulness. Actually when energy is consumed, we do not "consume" energy, but the available work. As available work is consumed, the quality of energy is degraded; the quantity of energy remains the same. Hence good energy saving practice strives to harness energy at the highest quality or temperature possible: that is to avoid unwanted degradation due to friction, or from large temperature or pressure drops, or through mixing of different temperature energy flows.

The following example illustrates that it is wasteful to burn fuels just to obtain low quality energy needed for low temperature process heat.[3] Consider for example two cases where electricity and steam or hot water or process heat are produced. In case A, electricity and heat are produced independently, and the combined efficiency of the process is 52%, and in case B where cogeneration approach is used, the efficiency is 85%. The data is shown in Figure 1-1.

From the data it appears that it is more efficient to first produce high pressure steam at a temperature of 500° C. The available work in the steam is used to drive a back pressure turbine, where it is converted to mechanical energy that drives an electric generator. The steam at the output of turbine at a temperature of 150 to 175 C is used to fulfill the thermal needs. With the scheme in case B, it is possible to convert roughly 30% of the quality energy in the fuel to electricity and 50% to useful low temperature heat. The conventional method of producing heat and electricity separately loses almost 50% of the energy content of the fuel. With cogeneration, it is possible to reduce these losses to perhaps only 20%.

In a cogeneration process, the amount of steam flowing to the condenser and the resulting heat rejected to the condenser are reduced when part of the steam is extracted for process heat. This approach provides improved cycle efficiency. As process extraction increases, cycle efficiency will continue improving to the point where all steam is extracted. This effect is shown in Figure 1-2,[4] which assumes that the extraction to process is at a point about halfway through the turbine expansion. The extraction pressure also affects the cycle efficiency.

Case A:  Separate production of electricity and heat
          (combined first law efficiency $\eta$ = 52%)

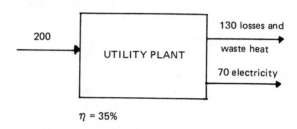

$\eta$ = 35%

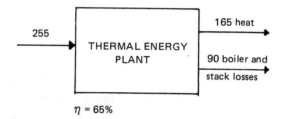

$\eta$ = 65%

Total fuel requirement for separate production is 455 units/h

Case B: Cogeneration of heat and electricity

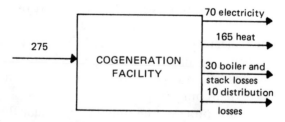

Total fuel requirement for cogeneration is 275 units/h
    (combined first law efficiency $\eta$ = 85%)

Rate of fuel savings is 180 units per hour, or 40% less than separate
production with same useful output

**FIGURE 1-1.  Typical Power Balance for Separate Heat and Electricity
          Production Compared with Cogeneration.**

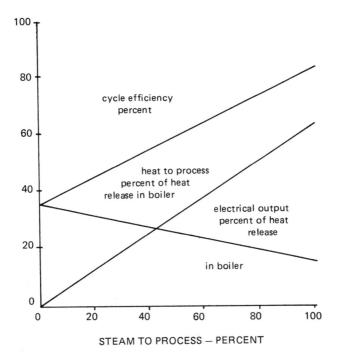

**FIGURE 1-2. Cycle Efficiency as a Function of Steam to Process.**

From an efficiency point of view, it is advantageous to have as low an extraction pressure as possible so that the steam produces maximum power before being extracted.

## COGENERATION TECHNOLOGIES

There are various energy conversion devices[5] that can be used in cogeneration facilities. These devices include the conventional and some newly developed energy conversion devices. One major consideration in selecting an appropriate energy conversion device is the ratio of electricity and steam which it produces. The ratio should closely match the electricity and steam demand of the anticipated energy market, otherwise the benefits from cogeneration will not be fully utilized. Another consideration for a cogeneration facility is its capability and flexibility to use various types of fuels in the event of fuel shortages and disruption of fuel supply.

There are two general concepts involved in cogeneration: topping cycle and bottoming cycle. In topping cycle, electricity is generated first and the waste heat in the form of exhausted steam is utilized for process steam or thermal energy needs. The energy conversion devices used in topping cycle are: diesel engines, gas turbines, combined gas cycles, steam boilers, and fuel cells.

In the bottoming cycle, the waste heat from an industrial plant is used to produce electricity. The energy conversion devices used in the bottoming cycle are: steam waste boilers, and organic Rankine cycle engines. The main characteristics and limitations of various energy conversion devices used in cogeneration facilities are summarized in Table 1-1.

A hypothetical 26 MW coal-fired, indirect-heated gas turbine cogeneration plant has been described,[6] including the plant arrangement, mode of operation, capital cost and operating economics. Another 100 MW coal gasification combined gas cycle demonstration project is underway at the "Cool Water" site of Southern California Edison Company.[7] This project will demonstrate a large coal gasification system. It will establish the environmental performance of the concept and viability of integrated operation with gas turbines, steam turbines, and other steam heat recovery equipment. In addition, it will also provide the technical and experience base needed for subsequent commercialization to industry and utility applications.

A 4.5 MW demonstration fuel-cell plant sponsored by the Department of Energy, the Electric Power Research Institute, and Consolidated Edison of New York[5] has been implemented in New York City. A molten carbonate fuel cell[8] integrated with a coal gasifier is one of the most promising coal based technologies for electric power.

Preliminary projections indicate efficiencies exceeding 60% as compared to 35% figure of a typical utility plant. The high electrical efficiency without need of a bottoming cycle and the quality of waste heat available serve as persuasive arguments for continued support of this technology and its early adoption once demonstrated. The above figure can be further improved if combined with bottoming cycle.

**TABLE 1-1. Characteristics of Various Energy Conversion Systems in Cogeneration Facilities**

| System | Capacity (MW) | Electicity-to-Steam Ratio | Type of Fuel | Suitability and Drawbacks |
|---|---|---|---|---|
| **Topping Cycles** | | | | |
| Diesel cogeneration | 0.5 - 25 | 400:1 | premium liquid fuels | industries where steam needs are minimal |
| Gas Turbines | 0.5 - 75 | 200:1 | natural gas, low Btu synthetic gas, light distillate oils, ethanol and methanol | excess electricity is produced. The facility should be able to sell the excess electricity. |
| Combined Gas Cycles | 1 - 150 | 150:1 | gas and liquid fuels for gas turbine, steam boilers can use solid, liquid, and gaseous fuels | requires transmission of electricity in some cases where all the electricity cannot be used by the industry |
| Conventional Rankine Cycle Extraction Turbine | 1 - 600 | 45 to 75:1 | greatest flexibility for fuel use | utilities favoring cogeneration |
| Back Pressure Turbines | 1 - 600 | 45 to 75:1 | greatest flexibility for fuel use | industries where steam is required as part of process heat |
| Fuel Cell | 1 - 150 | 300:1 | hydrogen and oxygen or air, excess electricity can be used to generate hydrogen and oxygen | suitable for facilities with high electric and low thermal demand, ideal in isolated facilities. |
| **Bottoming Cycles** | | | | |
| Steam Waste Boilers | 0.5 - 10 | — | high temperature heat that is wasted otherwise | brick kilns, glass furnaces, blast furnaces |
| Organic Rankine Engines | 0.5 - 1 | — | waste heat having temperature above 600 F | organic fluids are toxic and flammable thus hazardous |

The concept of power plants based on coal gasification is well on its way to commercialization through the "Cool Water Project," but its industrial applications will benefit as cogeneration, trigeneration or even "polygeneration" energy facilities are possible based on clean gas from coal as shown in Figure 1-3.

## CONSTRAINTS ON COGENERATION

Cogeneration is an attractive energy savings approach. Still, there are several obstacles to industrial and commercial cogeneration.[9]

- High cost of capital investment. Costs of cogeneration systems vary depending upon the size and the type of facility, but are high by any standard. 50-100 million dollars are typical costs for some types of systems. In tight economic conditions, industries do not have the necessary capital to install such facilities.

- Environmental concerns.

- Lack of restrictions on the use of oil and natural gas by utilities and power plants. In spite of the Power Plant and Industrial Fuel Use Act, there are several exemptions, where oil and natural gas are being used.

- Current low cost of electricity. Despite the rate increases of recent years, the cost of electricity still remains low for large industrial users due to the declining block rate structuring approach used by utilities.

- Restricted kwh revenue. The Federal Energy Regulatory Commission (FERC) has required utilities to purchase cogenerated industrial electricity, minimizing this obstacle, but the utilities pay a rate on an "avoided cost" basis.

- High back-up rates. Electric utilities have traditionally charged high rates to provide stand-by power. The FERC has ruled that electric utilities must apply the theory of load diversity in a non-discriminatory fashion to establish stand-by rates.

In the commercial and residential sectors, district heating is the only energy demand large enough to significantly accomodate the huge quantity of waste heat available from utilities. To some degree,

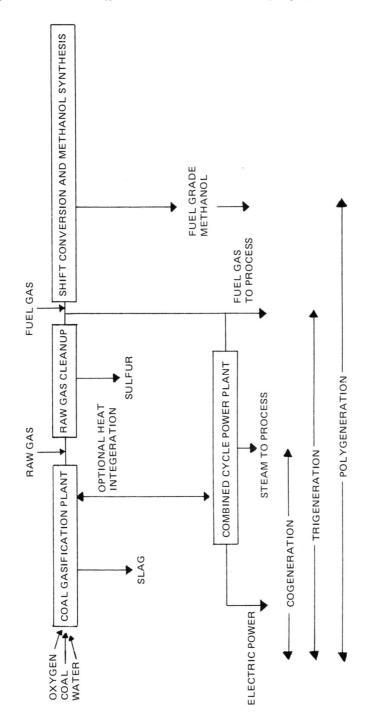

**FIGURE 1-3.** Concept of Trigeneration, Polygeneration Based on Coal Fueled Energy Conversion Systems

industrial waste heat is also a potential source. District heating can include both air-conditioning by the use of absorption chillers and heating of the buildings. The biggest drawbacks include logistics and capital costs:

- Steam or hot water transmission pipes from the power plant to the energy load centroid within the city.

- Distribution pipes which further carry the steam or hot water to individual buildings.

- Modification or replacement of heating and air-conditioning equipment in buildings to accomodate steam or hot water system.

- Because most utility power plants in U.S. have condensing turbines, they would have to be modified to enable steam extraction or replaced by extraction or back pressure turbines. All of the above improvements require huge capital investments at the part of the utilities, and also require planned development of cities.

## NUCLEAR COGENERATION

The concept of nuclear cogeneration has been widely discussed, and it is closer to reality in Europe than in the U.S. The Dow Chemical Company in Midland, Michigan, has built the first cogeneration nuclear power plant. The facility has been built by Bechtel Power Corporation for Consumers Power Company, and will soon be in operation.

The Dow Michigan Division is expected to fulfill its steam requirements from the Consumers Nuclear Power Plant. The steam will be cogenerated. High pressure steam will pass through the turbines to generate electricity for the network, and the medium pressure steam will be extracted for use by Dow plants. A portion of the medium pressure steam will pass through the Dow company turbines to produce electricity and low pressure steam for heating.

The dual role of a nuclear power plant has been viewed by many as an important way to conserve energy, reduce cost, and promote energy efficiency. The schematics of the nuclear cogeneration plant at Midland, Michigan, are shown in Figure 1-4. The concept of

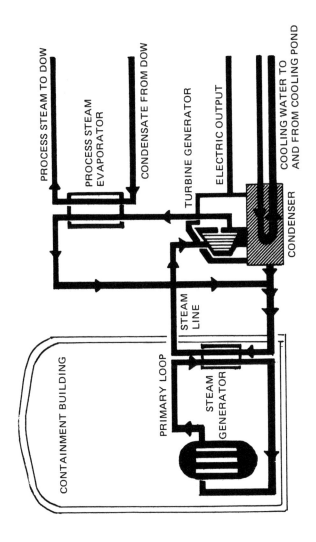

**FIGURE 1-4.** Dow Chemical Company Nuclear Cogeneration Power Plant[10]

cogeneration will also be beneficial in nuclear fusion energy when fusion energy becomes a reality.

## COGENERATION IN INDUSTRY

This section discusses the potential and concept of cogeneration technology in industry. Heat or thermal energy is used as dry heat in an industrial process or it is converted to steam. The equipment that makes the bulk of the energy used in an industrial system is called a prime mover. Steam and gas turbines, and diesel engines, have been used as prime movers by industry. Lately most of these prime movers have operated on natural gas and petroleum fuels due o their abundance and low price and less environmental problems. The conventional industrial operation in which thermal and electrical energy are generated independently is shown in Figure 1-5.[11]

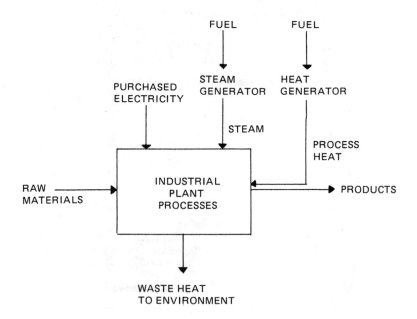

FIGURE 1-5. Conventional Industrial Process.

An alternate approach is that electrical and thermal energy can be cogenerated by using topping or bottoming cycles. In the topping cycle, the by-product steam or heat from a prime mover is used for plant processes. Waste heat from a plant can be recovered using bottoming cycles to generate electricity. The schematics of these two approaches are shown in Figure 1-6.[11] Both topping and bottoming cycles are significantly more fuel efficient than conventional systems that generate electric and thermal energy separately.

The potential for cogeneration in the chemical, petroleum refining, paper and pulp, textile, and food, and kindred product industries has been discussed in Energy Technology Review #29.[11] These industries are all among the ten largest energy users in the country.

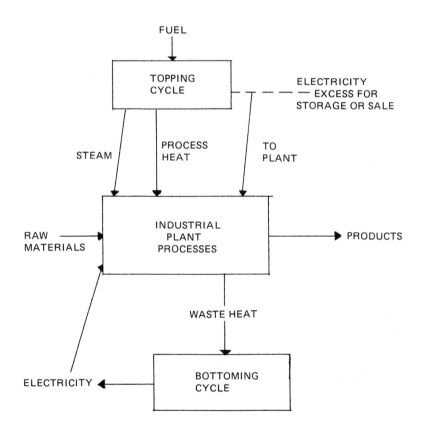

**FIGURE 1-6. Cogeneration Concept in an Industrial Process.**

Thus cogeneration becomes an attractive option as an energy savings and cost reduction approach. Investigations of these industries indicated that the managers had some common concerns that affected their enthusiasm for cogeneration:

- Cogeneration technology is not well known.

- Fuel supplies and prices are uncertain.

- Fuel flexibility is important, and cogeneration equipment needs to demonstrate this characteristic.

- Steam turbines are commonly used by these industries to meet their thermal and electric needs. However, steam turbines do not always satisfy electrical needs, so they have to buy back-up electricity.

- Gas turbines and diesels have the capability to meet electric demand better than steam turbines but they are expensive to buy and operate, and also require expensive and scarce liquid fuels.

- Excess electricity if sold can make diesels attractive economically, but there are technical and institutional problems, especially the fear that a plant manager might end up as a utility manager.

- Improved energy storage systems to store excess electricity or thermal energy are still not available.

- Cogeneration may require additional fuel which would result in increased emissions at the plant site, causing more environmental problems.

- For small to medium industries an electric grid may be the most reliable and flexible source of electricity.

- In industries where cogeneration has been adopted, 9 to 12% reductions in fuel use have been observed, but they had to invest large amounts of capital in equipment and face the problem of excess electricity. Thus smaller and cheaper cogeneration equipment may be adequate for their needs.

In spite of above concerns, the potential for adoption of cogeneration technology by industries exists and offers a significant promise of energy savings and cost reduction.

## TECHNICAL AND ECONOMIC FEASIBILITY

Current technologies plus expected improvements and increasing prices of energy make cogeneration technically and economically feasible. Some industries have been operating cogeneration plants for more than 20 years. The greatest near-term potential is the topping cycle. The potential of bottoming cycle cogeneration will continue to increase with time as the fuel costs keep on increasing. Burning of classified material and combustible municipal solid waste, utilizing fluidized bed combustion, combined gas cycles, polygeneration, and nuclear cogeneration are technically feasible and have a good potential for the future.

Cogeneration is economically feasible in certain industries where electric rates are high. Use of cogeneration can be easily justified for new and replacement facilities where good utilization can be experienced and sufficient electricity and thermal energy are required. In some cases the return on investment can be as high as 40% with a payback period of as little as two years. The data in Figure 1-7[12] indicates that if the return is 20%, the payback period is 5 years, and if the return is 10%, the payback period is 10 years.

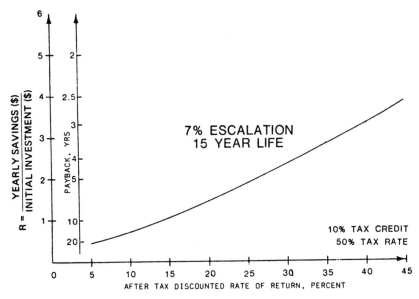

**FIGURE 1-7.** Economics of Cogeneration.

The industries groups that may be considered as prime candidates for cogeneration are: chemicals and allied products; primary metals; petroleum refining; paper and allied products; stone, clay, and glass products; and kindred products.

Some of the technical and economic constraints on increased development and implementation of cogeneration are summarized as follows:

- Variation in operating requirements.
- Lack of necessary load factors.
- Available technology.
- Seasonal demand for steam or process heat.
- Relatively low electricity prices.
- Cost of retrofitting.
- Availability of tax credits.
- Normal return on investment hurdle rates.
- Payback periods.
- Government action and inaction.
- Fuel price and availability.
- High capital investments and availability of capital.
- Equitable selling price for excess power.
- Air and water pollution standards.

## FUTURE OF COGENERATION

Cogeneration and district heating will stimulate and assist in bringing a change to alternate forms of energy. Cogeneration and district heating are readily adaptable to new forms of energy such as: geothermal, nuclear, solar, biomass, solid wastes, coal burning processes and other alternate energy sources. "Future prospects of cogeneration are excellent, and even currently available systems are cost competitive with conventional electric utility systems," the Department of Energy reported in 1978.

One of the goals of the present energy policy in the U.S. is to reduce future energy demand through energy efficiency and fuel substitution. Cogeneration has the potential to play a significant role in achieving this objective for the United States thereby increasing our national security and improving our economy.

### References

1 C.R. Cummings and R.T. Sperberg, "Cogeneration Energy Systems Assessment," *Proceedings of the 5th World Energy Engineering Congress*, Atlanta, GA, 133, 1982.

2 D. Yergin, *Conservation: The Key Energy Source in Energy Future*, Ballantine Book, New York, 198-201, 1980.

3 *Efficient Energy Management: Methods for Improved Commercial and Industrial Productivity*, Prentice-Hall Inc., 1983.

4 H. Stead, Jr., "Steam Turbines for Cogeneration," *Proceedings of the 16th Intersociety Energy Conversion Engineering Conference*, 2096, 1981.

5 R.T. Sheahan, *Alternate Energy Sources: A Strategy Planning Guide*, An Aspen Publication, 19-35, 1982.

6 R.W. Foster, "Coal Fired Air Turbine Cogeneration," *Proceedings of the 17th Intersociety Energy Conversion Engineering Conference*, 2103, 1982.

7 J.R.M. Alger, "Cool Water Demonstration Project and its Industrial Applications," *Proceedings of the 17th Intersociety Energy Conversion Engineering Conference*, 384, 1982.

8 J.B. O'Sullivan and L.G. Marionowski, "The Molten Carbonate Fuel Cell Technology in Cogeneration Applications," *Proceedings of the 17th Intersociety Energy Conversion Engineering Conference*, 374, 1982.

9 "Cogeneration Utility Systems," National Electrical Contractors Association, Index No. 302532, December, 1980.

10 R. Meador, *Cogeneration and District Heating*, Ann Arbor Science Publishers Inc., 141-153, 1981.

11 R. Noyes, *Cogeneration of Steam and Electrical Power*, Noyes Data Corporation, 1978.

12 D. Battles, "Technical and Economic Feasibility of Cogeneration," *Industrial Energy Managers' Sourcebook*, The Fairmont Press Inc., 384-391, 1981.

13 U.S. Department of Energy, Division of Fossil Fuel Utilization Report DOE-FFU-1703, 1978. *Cogeneration: Technical Concepts - Trends - Prospects.*

# CHAPTER 2
# Prospects for Industry

*K.M. Clark, D.E. Criner*

A prospective industrial cogenerator evaluates cogeneration in much the same way that it assesses other capital expenditures. Firms expect to receive a payback of their capital expenditures in a minimum period of time. The minimum after-tax payback period varies but typically may be three to five years, or less.

There are many possible variations in configuring cogeneration plants. A prevalent approach is to use a natural gas-fired combustion turbine with a waste heat recovery boiler. In an effort to identify those industries and geographical areas most conducive to cogeneration, we have examined the typical payback periods for combustion turbine cogeneration plants under a variety of conditions.

To perform this analysis, Burns & McDonnell's cogeneration investment analysis computer program, COMAC, was employed, along with baseline information developed in performing a number of cogeneration feasibility studies for industrial and utility clients. Analyses were performed for a variety of input assumptions to assess the profitability of cogeneration over a range of cogeneration plant sizes, electric rates, fuel costs, unit capacity factors, and future escalation rates for fuel prices and electric rates. Although this generic approach ignores site-specific factors, it has yielded important conclusions regarding the probable direction of future cogeneration activities.

## CAPITAL COST VARIATIONS

The initial capital cost is critical in assessing the economic feasibility of a prospective cogeneration project since the length of the pay-back period is essentially proportional to the initial capital cost.

Figure 2-1 shows the typical variation in initial capital cost, $/kW, as a function of cogeneration unit size. These capital costs include the combustion turbine, heat recovery steam generator, installation, spare parts, initial inventories, interest during construction, engineering and legal fees, and owner's overhead incurred during construction.

It may be seen from Figure 2-1 that there are economies of scale that affect smaller combustion turbine cogeneration plants, particularly below about 20 MW. In addition to higher unit capital costs ($/kW), smaller units also incur proportionately higher operation and maintenance costs. Figure 2-1 suggests that combustion turbine cogeneration units smaller than about 20 MW will in general, encounter economic obstacles. Each case must be analyzed on its own merits and parameters, however.

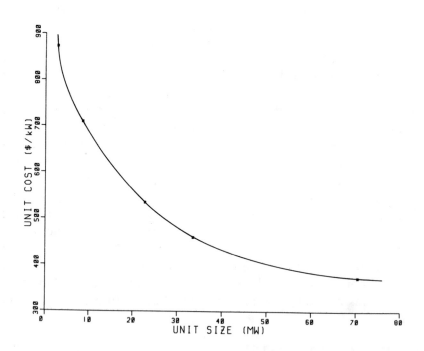

**FIGURE 2-1. Capital Cost vs. Unit Size for Sample Cycle Gas Turbine with Waste Heat Recovery Boiler.**

## SENSIVITY TO FUEL AND ELECTRIC RATES

Next to capital cost, the critical factors involved in determining the economics of a cogeneration unit are the fuel cost and the electrical rates faced by the prospective cogenerator. Figures 2-2 through 2-6 show how the projected after-tax payback period varies with respect to variations in fuel cost ($/MMBtu) and in electric rates (cents/kWh).

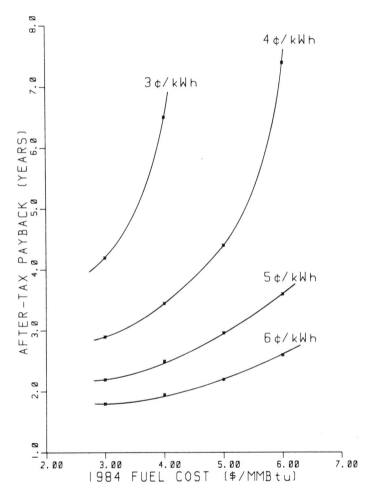

**FIGURE 2-2. After-Tax Payback Period vs. Fuel Cost for 70-MW Gas Turbine with Waste Heat Recovery Boiler.**

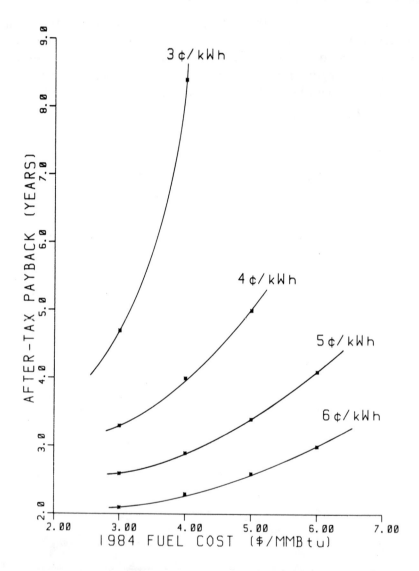

**FIGURE 2-3.  After-Tax Payback Period vs. Fuel Cost for 33-MW Gas Turbine with Waste Heat Recovery Boiler.**

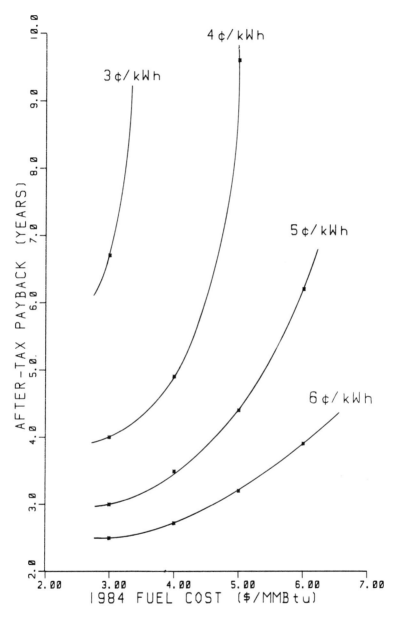

FIGURE 2-4. After-Tax Payback Period vs. Fuel Cost for 23-MW
Gas Turbine with Waste Heat Recovery Boiler.

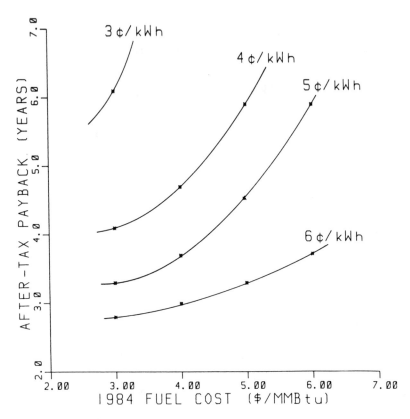

**FIGURE 2-5. After-Tax Payback Period vs. Fuel Cost for 8.5 MW
Gas Turbine with Waste Heat Recovery Boiler.**

Each of these figures is for a different size cogeneration unit—
from 70 MW down to 2.8 MW. Each plot is based upon COMAC
computer runs using an annual escalation rate of 5 percent for all
costs, a 1986 in-service year for the cogeneration unit, manufacturers' equipment performance data, variable and fixed operation and
maintenance costs derived from Burns & McDonnell's experience, an
income tax rate of 50 percent, a 10 percent investment tax credit,
and an 85 percent unit capacity factor. The value of the steam
generated by the cogeneration unit was determined using the price
of natural gas, an assumed 80 percent efficiency for a conventional
boiler, and allowances for the operation and maintenance of a conventional boiler.

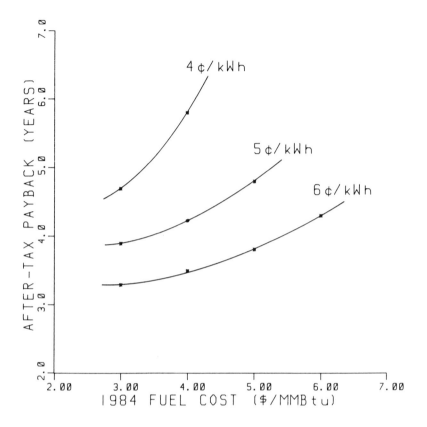

**FIGURE 2-6.** After-Tax Payback Period vs. Fuel Cost for 2.8 MW
Gas Turbine with Waste Heat Recovery Boiler.

In reviewing Figures 2-2 through 2-6, it is apparent that the larger units have shorter payback periods than smaller units. The most favorable economics are projected for the 70-MW unit (Figure 2-2).

In many areas of the country, a typical price for natural gas may be $4/MMBtu and an industrial rate for electricity may be 4 cents/ kWh (including both energy and demand components). Under such typical assumptions, the after-tax payback period for a 70-MW unit (Figure 2-2) is seen to be 3.5 years. This would be regarded as an acceptable payback by many industrial organizations.

However, for the same fuel cost and electric rate assumptions, the payback periods for the smaller units became less attractive as summarized below:

| Unit Size | | After-Tax |
| --- | --- | --- |
| Electrical (MW) | Steam (1000 lb/h) | Payback (years) |
| 70 | 375 | 3.5 |
| 33 | 193 | 4.0 |
| 23 | 142 | 4.9 |
| 8.5 | 43 | 4.7 |
| 2.8 | 19 | 5.8 |

## CAPACITY FACTOR

For the analyses presented in Figures 2-2 through 2-6, a capacity factor of 85 percent was assumed for each cogeneration unit. Higher capacity factors, although achievable, may be optimistic considering periodic outages that will occur for cogeneration plant maintenance, regular plant shutdowns that are scheduled in some industries, and hourly, daily, and weekly variations in steam demand that often occur.

Figure 2-7 shows the sensitivity of payback period to lower capacity factors. (This particular curve was prepared for the 33-MW unit previously depicted in Figure 2-3.) As the assumed capacity factor decreases, the economic payback period increases dramatically.

It may be concluded that ideal cogeneration applications should use the full output of the cogeneration unit on a 24-hour/day, 7-day/week basis. Lower usage factors will encounter more difficult economic hurdles. Therefore, cogeneration units serving seasonal thermal loads (such as for space heating) will, in many cases, be difficult to justify. A similar conclusion may be drawn for process applications that only operate one or two shifts per day.

## ESCALATION RATES

Figures 2-2 through 2-6 are based upon a 5-percent annual escalation rate applied uniformly to all costs, including fuel and electricity. If electric rates escalate more rapidly than fuel, then the payback

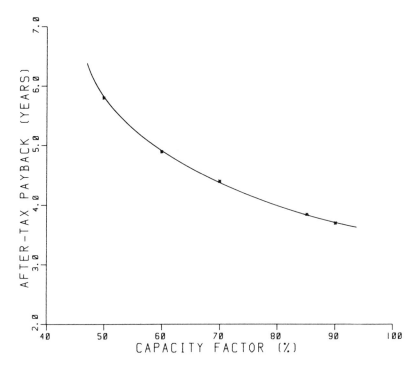

**FIGURE 2-7. Payback Period vs. Capacity Factor for 33 MW Unit.**

periods become shorter. This effect is displayed in Figure 2-8 (which is based on the same parameters as the 33-MW unit shown previously in Figure 2-3).

If one assumes that fuel and electricity escalate at different rates, the evaluation may distort the economic analysis of cogeneration. If revenues (or savings) are projected to escalate at a higher rate than costs, even a poor investment decision may appear attractive.

Therefore, in evaluating cogeneration projects, higher escalation rates for electricity than for fuel should not be assumed without justification. This is particularly important if electricity from the prospective cogeneration unit is to be sold to the utility at avoided costs to be determined in the future; such avoided costs can drop precipitously—even as the rates the utility charges its customers increase.

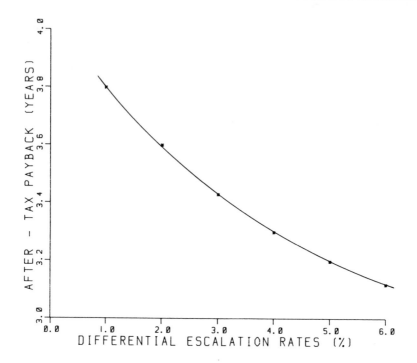

**FIGURE 2-8. Payback Period vs. Differential Escalation Rates (Electricity and Fuel).**

## PLANT RELIABILITY AND BACKUP

The economics of cogeneration will be impacted by the reliability that is achieved for cogeneration units. First, if a cogeneration unit proves to be unreliable, its capacity factor will deteriorate, which, in turn, will adversely affect the economic payback for the investment. Also, poor unit reliability is associated with high maintenance costs and lengthy initial commissioning periods—either of which can eliminate the profitability originally perceived for a cogeneration project.

Perhaps even more critical is the potential disruption to a cogenerator's basic processes when an unplanned outage of the cogeneration unit occurs. To minimize such impact, cogeneration plants must be carefully engineered to provide appropriate backup steam supplies, recognizing that a combustion turbine will not achieve the same reliability as a conventionally fired boiler. Also, a combustion turbine plant is more complex than a conventional boiler, thus requiring a more sophisticated maintenance program.

Experience shows that achieving high reliability for a congeneration plant requires not only a reliable combustion turbine, but also requires careful system engineering of balance-of-plant auxiliaries and support systems and an effective maintenance program.

## THE QUESTION OF CAPACITY CREDITS
### (The Utility Perspective)

Under the Public Utility Policies Act (PURPA), electric utilities are required to interconnect with cogenerators and, if requested by the cogenerator, to purchase electric power on the basis of the utility's avoided costs. The avoided costs include an energy component (cents/kWh) and, in some cases, a demand component ($/kW-mo). One of the challenging aspects of cogeneration economics has been the determination of the capacity credit to be paid to cogenerators.

### Excess Generating Capacity

In many areas of the United States, there is a surplus of electric generating capacity. As a result, many utilities offer no capacity credits to cogenerators since the capacity installed by a cogenerator merely adds to the utility's surplus capacity.

However, regulatory commissions in several states are attempting to encourage cogeneration as a means of energy conservation and as a response to the influence of cogeneration groups. As a result, some utilities are offering capacity credits for cogeneration—even in the face of surplus generating capacity—on the basis that new cogeneration units will ultimately defer installation of new, utility-owned units in the future. For example, assume that a utility expects to need additional capacity in 1992 that it is expected to cost $2,000/kW. Then P, the present (1984) value of this future capacity, is given by:

$$P = \frac{\$2000}{(1+i)^8} \qquad \text{(Equation 1)}$$

where i is the appropriate discount factor. This present value, P, may then be used to derive a levelized capacity credit to be paid to new cogenerators. This capacity credit, C, ($/kW-yr) is given by:

$$C = P \frac{i}{(1+i)^n - 1} + i \qquad \text{(Equation 2)}$$

where n is the number of years assumed for levelizing.

Variations of this approach for establishing capacity credits are being implemented by several utilities.

## Utility Concerns

In spite of such theoretical approaches available for computing capacity credits, there are several utility concerns that are difficult to resolve—particularly the uncertainty regarding the longevity of industrial cogeneration units.

Industrial cogenerators often expect a short payback period for their investment in cogeneration facilities because their industrial plants can be adversely affected by future market conditions, foreign competition, technology changes, and labor costs, Therefore, an industrial facility that is now profitable could face closure or relocation several years from now. What if the utility pays a cogenerator a levelized capacity credit in anticipation of deferring new generating capacity that would otherwise be needed in 1992, but the cogenerator ceases operation in 1991?

One solution to this concern is for the cogenerator to enter into a long-term contract to provide capacity to the utility. However, many cogenerators are unwilling to enter into a sufficiently long-term agreement to be factored into the utility's long-range system planning.

## In-Plant Use

Because of the uncertainty in capacity credits to be paid to cogenerators, much of the current cogeneration activity in the United States is by industrial firms that can use the electric power in their own plants, thus displacing power that would otherwise be purchased from the local utility. Since the cogenerator does not sell power to the utility, the question of capacity credits is avoided.

Also, if the cogenerator does not choose to sell power to the utility, contractual negotiations between the cogenerator and the utility are simplified with the result that a project may be implemented more quickly.

## Historical Perspective

There are certain parallels between current cogeneration developments and the development of "total energy" systems several decades ago. During the 1960s, there was much interest in installation of total energy systems supplying all of the electrical and thermal requirements for an industrial owner. (The term "total energy system" was used before the word "cogeneration" was invented.) Some of the total energy systems that were installed were later abandoned because practical operating difficulties were encountered.

It is inevitable that there will be similar unsuccessful experiences in the current wave of cogeneration activity. A lesson to be learned is that prospective cogenerators should carefully analyze the economics of proposed cogeneration systems and ensure that the systems are engineered to facilitate reliable operation and maintenance. Successful cogenerators also should develop effective capabilities for routine operation and maintenance.

## Financing Alternatives

Cogeneration projects involving third-party financing or ownership have received much attention. Third-party financing arrangements may permit the installation of a cogeneration plant with minimum capital outlay by the cogenerator. However, in the author's experience, such innovative financing arrangements sometimes become entangled by financial and legal questions, tax issues, and lengthy negotiations concerning the responsibility for the inherent risks of the project. In some cases, innovative financing and ownership arrangements have been employed as a "gimmick" to promote cogeneration projects for owners that have limited staffs or are otherwise ill-equipped to evaluate or implement the technical aspects of a cogeneration project.

Although third-party financing and ownership arrangements will continue to serve an important role in cogeneration development,

there are many less-publicized cogeneration projects being implemented by well-established industrial companies with internally
generated funds. Many of these companies have sufficient technical
expertise to make their own evaluation of cogeneration projects and
wish to avoid the complications involved in third-party arrangements.


## SUMMARY AND CONCLUSIONS

On the basis of these studies and other experience, the authors
draw the following conclusions:

1. There are significant economies of scale that apply to combustion turbine cogeneration units. This could discourage the
   installation of smaller combustion turbine units depending on
   the application.

2. The economics of cogeneration units are sensitive to fuel prices
   and electric rates—which obviously vary from region to region.
   For typical fuel and electric costs prevalent in many areas of
   the United States, the larger combustion turbine cogeneration
   units have attractive after-tax payback periods.

3. Cogeneration economics deteriorate rapidly as the unit capacity
   factor decreases. This will discourage cogeneration units for
   applications that do not require process heat on an essentially
   continuous, around-the-clock basis.

4. If electric rates escalate faster than fuel prices (particularly
   natural gas), then the economic feasibility of cogeneration will
   be enhanced. However, assuming such differential escalation
   rates may distort the projected payback period and, therefore,
   any such assumption should be carefully justified.

5. There are continuing uncertainties involved in establishing
   capacity credits for electric power sold by cogenerators to
   utilities. In some areas, this will tend to discourage prospective
   cogenerators that cannot use the total electrical output of a
   cogeneration unit.

6. Innovative financing and ownership concepts will continue to play a prominent role but can add complications and potential project delays compared with conventional arrangements.

7. Cogeneration projects that "seem good on paper" can encounter practical difficulties that eliminate the profitability of the project. To minimize the prospect for such problems, thorough engineering of each cogeneration project will be required.

In general, the authors believe that cogeneration will continue to prove attractive in many applications. However, the underlying economics of cogeneration often do not capture as much attention as imaginative and complex financing packages, overly simplified evaluations of some cogeneration projects, the novelty of the Public Utility Regulatory Policies Act (PURPA), and enthusiasm to conserve energy.

In the future, some of these factors probably will receive relatively less attention—with a resultant "shake-out" in the cogeneration industry and a maturing perception of cogeneration by industry, utilities, consultants, and suppliers. Although this appears inevitable, the authors believe the underlying economics of cogeneration are sound and that knowledgeable industries and utilities will continue to implement cogeneration for appropriate applications.

# CHAPTER 3

# Economic Attractiveness of Cogeneration Technologies for the Industrial Steam User

*M. Whiting, Jr., G.L. Decker*

In recent years new factors have emerged to make cogeneration increasingly attractive. Most important has been the enormous escalation of energy prices over the last decade, which has made energy efficiency an important economic factor for most industrial energy users. The passage of the Public Utilities Regulatory Policy Act (PURPA) in 1978 removed most of the institutional constraints which had previously hampered the development of cogeneration.

The high cost of new central station electric capacity has caused some utilities to encourage cogeneration as a way for them to avoid having to raise capital for new plants. In addition, cogeneration technology has continued to improve as more efficient and lower cost systems have been developed. The overall effect of these factors has been to stimulate interest in cogeneration as a means to lower energy costs and to achieve national energy conservation goals.

The fundamental requirement in determining if cogeneration is feasible is the existence of substantial and fairly constant demand for heat, because the efficiency of a cogeneration system is dependent on the ability to use both the electric and the thermal outputs. Without the efficiency of cogeneration, it is very difficult for an industrial power producer to generate electricity more economically than large utility plants, which often use low-cost coal or nuclear fuel. Cogeneration is most favorable in areas with high electric rates, since the system economics are most dependent on the value of the electric output.

## ADVANTAGES OF COGENERATION

The great interest in cogeneration results from the manifold advantages it presents to industrial energy users.

- The most significant benefit of cogeneration is its high efficiency compared to the separate production of electricity and heat. In Figure 3-1 a comparison is made between the efficiency of energy utilization in a cogeneration system and in a conventional power plant.[1] By using otherwise wasted heat, cogeneration can produce electricity and steam at a much lower cost than making them separately. These cost savings resulting from increased energy efficiency are the driving force for cogeneration.

- The capital cost of cogeneration systems is usually lower than that of new utility central station capacity. Gas turbine combined cycle cogeneration systems are at least 50% less expensive on a $/kW basis than new coal or nuclear power plants.

- Cogeneration is an old and proven technology that has been around since before the turn of the century. It is well understood and very reliable.

- Despite its longevity, cogeneration technology continues to be improved as new systems with higher efficiencies and lower capital costs are developed. By the late 1980's substantial improvements will be realized with the commercial introduction of gas turbines with higher firing temperatures and fuel cells.

- Because of its high energy efficiency, cogeneration also presents environmental advantages, since less fuel is burned to produce the same energy output.

## DISADVANTAGES OF COGENERATION

Despite its many advantages there are a number of drawbacks to cogeneration which must be carefully considered before making a commitment to a cogeneration project.

- Cogeneration systems are highly complex to design, install, and operate. It cannot be done from a textbook. There is no substitute for in-depth, hands-on knowledge that can only be gained from many years of experience in working with cogeneration systems.

- Because electricity and steam are produced in fixed or nearly fixed ratios, cogeneration systems impose new constraints on plant operations. To achieve the maximum efficiency, production plants must coordinate their operation with the utility system to a greater extent than would otherwise be required.

- For the majority of companies that presently are not cogenerators, there is often a reluctance to branch into the electricity generating business, which is new to them and with which they have little or no experience.

- Cogeneration systems require a substantial capital investment which many companies are unwilling to commit to a project that does not increase their production capacity, even if the economics are highly favorable. In addition, a large commitment of other corporate resources is required, which many companies would rather devote to something more central to the operation of the organization.

- The economics of cogeneration projects are very dependent on electric utility rates and fuel prices, both of which can be highly upredictable. Cogeneration projects that depend on selling large amounts of power to the electric utility at avoided cost are particularly vulnerable, if there is no long-term contract which specifies the buyback rate.

- Despite the passage of PURPA and its recent upholding by the Supreme Court, some electric utilities remain opposed to cogeneration, which can complicate the development of a cogeneration project.

## STEAM SUPPLY ALTERNATIVES

Many industrial steam users today are faced with a difficult dilemma. They produce steam in old, inefficient boilers which were designed and built in the pre-1973 era of cheap energy. High energy prices have made these boilers obsolete and may threaten the position of these companies in world markets. To remain competitive, these companies must reduce their energy costs. In many cases cogeneration can provide them with a cheaper source of steam, resulting in savings that are directly reflected in higher profits.

In Table 3-1 conventional steam boilers are compared to the steam turbine and gas turbine combined cycle cogeneration systems, each having a 150 psig steam output of 100,000 lb/hr. This table demonstrates the benefit of cogeneration: producing electricity on top of the steam demand.

The steam turbine system generates 5.5 MW, and the gas turbine combined cycle system 25.6 MW. If it is assumed that only the incremental fuel usage is allocated to the generation of electricity, it requires only 4280 Btu/kWh for cogenerated power from the steam turbine system and 5080 Btu/kWh from the combined cycle system. These values compare extremely favorably with the average electric utility heat rate of 10,500 Btu/kWh, and it is on this basis that cogeneration is so attractive.

While the incremental electric generating efficiency of the steam system is better, the gas turbine combined cycle system has roughly five times the electric output for the same steam output. Since energy is in the form of heat, the output of the combined cycle system is much more valuable than that of the steam turbine system.

This is reflected in the Second Law efficiency found in Table 3-1, a measure which takes into account the quality as well as the quantity of the energy. As a result of this overall higher quality output, the combined cycle system has the greatest potential fuel savings, a 35% reduction compared to the separate production of electricity and steam.

Hence, the combined cycle system is the most attractive cogeneration system available today, as long as the electricity generated can be used or sold at an attractive rate.

In addition to its high efficiency and fuel savings potential, the combined cycle system offers several other advantages. It has a low capital cost, as little as $400/kW for large installations. The economies of scale are relatively small, so gas turbine cogeneration systems can be attractive in sizes as small as a few megawatts.

The lead time required to place a combined cycle system in operation can be as short as two years. This technology is well established and reliable. The first gas turbine for power generation was installed in 1949, and combined cycle systems have experienced availabilities in excess of 98% over many years of operation.

Finally, although combined cycle systems today must burn either natural gas or distillate oil, they can readily be converted to run on low or medium-Btu gas, when coal gasification becomes economic in the future. Thus over the long term, coal is available as a back-stop for the combined cycle system against future escalation of oil and gas prices.

There are, however, a few disadvantages of the gas turbine combined cycle cogeneration system. The most significant one is the requirement for premium fuels, namely natural gas, distillate oil, or coal gas. These fuels are expensive and have been subject to supply interruptions in some cases in the past.

Since such a large percentage of the output is in the form of electricity, the economics of combined cycle systems are highly dependent on the electric rate, either the price of electricity from the utility or the avoided cost. In addition, combined cycle systems usually generate electricity in excess of an industrial user's needs, so surplus power must often be sold to the electric utility, introducing added complications to the project.

## ECONOMIC EVALUATION

Its technical merits aside, the ultimate success of cogeneration depends on whether it makes sense economically. The best way to determine the attractiveness of a cogeneration project is to calculate the after-tax internal rate of return, which is equivalent to the compounded interest rate that a project earns. If the internal rate of return equals or exceeds the current market rate for investments of similar risk, the project is attractive enough to be financed.

To calculate the internal rate of return, the annual after-tax cash flow over a period of several years must be determined. To do this, several factors must be known or calculated, including the following:

**Revenues:**
Electricity (E)
Steam (S)

**Costs:**
  Fuel (F)
  Other (O&M, insurance, state and local taxes (C)
    Debt service [principal (P) and interest (I)]
**Tax benefits:**
  Depreciation (D)
  Investment tax credit (T)

Once these factors are known, the after-tax cash flow can be calculated according to the following equation (assuming a corporate tax rate of 46%):

$$\text{Cash Flow} = (1 - 0.46) * (E + S - F - C - I - D) - P + D + T$$

To determine the internal rate of return for the project, after-tax cash flows must be calculated for each of several years by applying the appropriate escalation factors to each of the above revenues and costs. The annual cash flows can then be entered into a computer or calculator, which can readily perform the calculation.

In Table 3-2 such an analysis is shown over a 10-year period for a gas turbine combined cycle cogeneration plant. This table assumes modest energy prices: $0.04/kWh electricity, $4.00/MMBtu fuel, and $5.35/1000 lbs steam. A 2% real escalation rate was applied to electricity and a 4% real escalation rate to fuel and steam.

Despite these conservative price assumptions, the after-tax internal rate of return for this cogeneration project is 30.2%. In today's financial markets investors might only demand after-tax rates of return of about 20% for cogeneration projects, which have relatively low levels of risk. Thus, under the above assumptions, cogeneration is extremely attractive.

If a cogeneration project is examined from the standpoint of the industrial steam user, the primary parameter of interest is the price of the cogenerated steam. Table 3-2 used a steam price equivalent to the fuel cost of steam from an 80% efficient boiler.

Another way to analyze the economics of the project is to assume that the equity investors in the project require a 20% after-tax return and to adjust the price of the steam to achieve that return. In effect, this approach allocates to the industrial steam user any excess profitability over that required by the investors.

In Table 3-3 such an analysis is shown, using the same assumptions as above except that the steam price was fixed to give the investors a 20% after-tax internal rate of return. The result is a steam price of $4.44/1000 lbs, or a 17% discount below the fuel cost of the steam with the investors achieving the required 20% rate of return.

In reality the true savings to the industrial steam user are even greater than this, since by purchasing cogenerated steam he also avoids having to pay operating and maintenance costs for conventional steam boilers, which typically are about $1.00/1000 lbs of steam. Thus the savings from buying cogenerated steam under these assumptions total about 31% compared to the cost of steam from a conventional steam boiler.

To fully understand the economic attractiveness of a cogenerated project, it is important to perform sensitivity analyses for the most significant or the most unpredictable parameters in the economic evaluation. For a combined cycle cogeneration plant the most significant variable in determining its economic attractiveness is the electric rate.

In addition, the fuel price and the capital cost are also important. The sensitivity of the steam cost savings (as described in two paragraphs directly above) to these parameters is shown in Figures 3-2, 3-3, and 3-4. These graphs highlight the great effect that the electric rate has on the economics of cogeneration projects. If the electric rate is increased from $0.04/kWh, the steam cost savings to the industrial user can be increased from 31% to 68%, while the investors continue to achieve a 20% rate of return. On the other hand, the fuel cost and the capital cost have the opposite effect, but to a lesser degree.

## THIRD-PARTY FINANCING

Many industrial steam users would like to realize the benefits of cogeneration without making a large capital investment or having responsibility for operating the plant. They would like to purchase cogenerated steam over-the-fence, allowing them to obtain steam directly without having to bother with buying fuel and converting it into steam in a boiler.

Third-party financing enables them to do exactly that: to obtain a low-cost supply of process steam at no capital cost to the company.

In addition, the financing can be done entirely off-balance sheet, and thus not affect the company's credit rating.

The basic requirement for organizing third-party financing is to have solid, long-term contracts for the fuel input and the steam and electricity outputs. Financial institutions will commit equity and debt funds to the cogeneration project based on the strength of these contracts. In addition, a cogeneration project must be structured with well-contained risks. The following major risk factors must be addressed in organizing the financing:

- *Construction risks*: whether the project will be completed on time and within budget;

- *Technical risks:* whether the equipment will perform to design specifications and with the expected operating and maintenance costs;

- *Fuel input:* whether fuel will be available in the quantity required and at an acceptable price over the duration of the project;

- *Market for output:* whether there is an assured market for the full electric and steam output, at what prices, and how the long-term prospects of the steam user's business might affect his ability to meet contractual obligations as a result of changes in the market for his products;

- *Financing risks:* whether the project will generate sufficient revenues to meet fixed debt obligations over the duration of the project and how changing interest rates will affect the project.

A properly structured cogeneration project will allocate each of these and other risks to the party best able to handle them. For example, the risk of suboptimal equipment operation is often best addressed by performance guarantees from the equipment vendor, who has the most control over that aspect of the system and has the most experience in dealing with those problems.

Another key facet of a cogeneration project is the existence of a project developer who sponsors the project and brings together the many parties involved. This developer could be any one of a number of types of firms: the fuel supplier, the industrial steam user, the electric utility, the equipment supplier, financial institutions, or entrepreneurs whose business is developing cogeneration projects.

Whoever it is, the developer must be able to devote a great deal of time to the project and to absorb large front-end costs without any assurance of recovering the investment. Cogeneration projects can take as long as four years to organize and can require an up-front investment of several million dollars. Without a dedicated and resourceful developer, a complex, third-party-financed cogeneration project is unlikely to ever get off the ground.

## CONCLUSION

Cogeneration, and the gas turbine combined cycle system in particular, offer the potential for substantial cost savings to industrial energy users. The electricity and fuel costs that exist today create great opportunities for cogeneration in many areas of the United States.

As more electric utilities bring on line expensive new coal or nuclear generating capacity, the differential between fuel and electricity costs will continue to widen, further enhancing the attractiveness of cogeneration.

The industrial steam user can obtain the benefits of low-cost cogenerated steam by cogenerating himself or by purchasing steam from a third-party owned and operated plant, and thereby avoid the capital expenditure and operating responsibility for the project. However it is done, cogeneration is one of the best ways available to substantially reduce energy costs and to improve an energy-intensive company's position in highly competitive world markets.

### References

[1] Dow Chemical Co., et al., "Energy Industrial Center Study," National Science Foundation, 1975.

[2] Arkansas Power & Light, "Cogeneration: A Threat or an Opportunity for the Electric Utility Industry," Coal Technology 1981 Conference.

[3] Marc H. Ross and Robert H. Williams, "Industrial Cogeneration: Making Electricity with Half the Fuel," chapter 10 in "Our Energy: Regaining Control," 1981.

**TABLE 3-1. Steam Supply Alternatives**

| | Standard Boiler | Cogeneration | |
| --- | --- | --- | --- |
| | | Steam Turbine | Gas Turbine Combined Cycle |
| 150# steam output (lb/hr) | 100,000 | 100,000 | 100,000 |
| Natural gas usage (MCF/hr) | 130 | 153 | 257 |
| Byproduce power (MW) | – | 5.5 | 25.6 |
| Power: Steam ratio (kW/100 lb/hr) | – | 55 | 256 |
| Electric output (% of fuel input) | – | 12 | 33 |
| First Law efficiency (%) | 80 | 80 | 73 |
| Second Law efficiency (%) | 25 | 38 | 52 |
| Cogeneration fuel savings (%) | – | 10 | 35 |

**Bases:**

150# saturated steam @ 1070 Btu/lb (80% efficient boiler) Electricity @ 3413 Btu/kWh.

Steam turbine inlet steam conditions: 1300#, 900 deg. F.

Combined cycle values based on G.E. MS-7001 (E) gas turbine normalized to 100,000 ln/hr

150# steam output (from reference 2).

First Law efficiency and Second Law efficiency from reference 3).

Cogeneration fuel savings (from reference 2).

## TABLE 3-2. Economic Evaluation of Combined Cycle Cogeneration

| Year | Revenue Power | Steam | Fuel Cost | Other Costs | Depreciation | ITC | Debt Int. | Service Prin. | After-Tax Cash Flow |
|---|---|---|---|---|---|---|---|---|---|
| 0 | | | | | | | | | -6.41 |
| 1 | 8.08 | 4.22 | 8.44 | 0.63 | 1.77 | 0.96 | 0.77 | 0.37 | 2.69 |
| 2 | 8.73 | 4.64 | 9.28 | 0.67 | 2.36 | | 0.73 | 0.41 | 2.10 |
| 3 | 9.43 | 5.10 | 10.21 | 0.71 | 2.23 | | 0.68 | 0.46 | 2.10 |
| 4 | 10.18 | 5.61 | 11.23 | 0.75 | 2.20 | | 0.62 | 0.51 | 2.14 |
| 5 | 11.00 | 6.18 | 12.35 | 0.79 | 2.17 | | 0.56 | 0.57 | 3.19 |
| 6 | 11.88 | 6.79 | 13.59 | 0.84 | 0.19 | | 0.49 | 0.64 | 1.19 |
| 7 | 12.83 | 7.47 | 14.94 | 0.89 | 0.19 | | 0.41 | 0.72 | 1.24 |
| 8 | 13.85 | 8.22 | 16.44 | 0.95 | 0.19 | | 0.33 | 0.81 | 1.30 |
| 9 | 14.96 | 9.04 | 18.08 | 1.00 | 0.19 | | 0.23 | 0.90 | 1.35 |
| 10 | 16.16 | 9.95 | 19.89 | 1.06 | 0.19 | | 0.12 | 1.01 | 1.40 |

After-tax internal rate of return = 30.2%

**Assumptions:**

| | | |
|---|---|---|
| Electric output | = | 32.7% of fuel input |
| Steam output | = | 40.0% of fuel input |
| Steam demand | = | 100,000 lb/hr 150# saturated steam |
| Electric output | = | 25.6 MW |
| Installed cost | = | $12.82 MM ($500/kW) |
| Capacity factor | = | 90% |
| Electric rate | = | $0.04/kWh |
| Fuel cost | = | $4.00/MMBtu |
| Steam price | = | $5.35/1000 lbs (100% of fuel cost) |
| General inflation rate | = | 6% |
| Electricity escalation rate | = | 8% |

*(continued)*

Fuel escalation rate = 10%
Steam escalation rate = 10%
Debt: equity ratio = 50:50
Interest rate = 12%
Ten-year project for economic evaluation purposes
Electric rate is an average value for all power sold
Installed cost includes equipment cost, and project fees
75% of investment receives 5-year accelerated depreciation and 10% ITC
25% of investment receives 15-year depreciation and no ITC
Other costs include state and local taxes, insurance, and O&M cost
150# saturated steam valued at 1070 Btu/lb
Electricity valued at 3413 Btu/kWh

(end)

TABLE 3-3. Economic Evaluation of Combined Cycle Cogeneration with Discounted Steam Price

| Year | Revenue Power | Steam | Fuel Cost | Other Costs | Depreciation | ITC | Debt Int. | Service Prin. | After-Tax Cash Flow |
|---|---|---|---|---|---|---|---|---|---|
| 0 | | | | | | | | | -6.41 |
| 1 | 8.08 | 3.50 | 8.44 | 0.63 | 1.77 | 0.96 | 0.77 | 0.37 | 2.36 |
| 2 | 8.73 | 3.85 | 9.28 | 0.67 | 2.36 | | 0.73 | 0.41 | 1.74 |
| 3 | 9.43 | 4.24 | 10.21 | 0.71 | 2.23 | | 0.68 | 0.46 | 1.70 |
| 4 | 10.18 | 4.66 | 11.23 | 0.75 | 2.20 | | 0.62 | 0.51 | 1.71 |
| 5 | 11.00 | 5.13 | 12.35 | 0.79 | 2.17 | | 0.56 | 0.57 | 1.71 |
| 6 | 11.88 | 5.64 | 13.59 | 0.84 | 0.19 | | 0.49 | 0.64 | 0.65 |
| 7 | 12.83 | 6.20 | 14.94 | 0.89 | 0.19 | | 0.41 | 0.72 | 0.66 |
| 8 | 13.85 | 6.82 | 16.44 | 0.95 | 0.19 | | 0.33 | 0.81 | 0.66 |
| 9 | 14.96 | 7.50 | 18.08 | 1.00 | 0.19 | | 0.23 | 0.90 | 0.65 |
| 10 | 16.16 | 8.25 | 19.89 | 1.06 | 0.19 | | 0.12 | 1.01 | 0.63 |

(continued)

*(end)*

After-tax internal rate of return = 20.3%

**Assumptions:**

| | | |
|---|---|---|
| Electric output | = | 32.7% of fuel input |
| Steam output | = | 40.0% of fuel input |
| Steam demand | = | 100,000 lb/hr 150# saturated steam |
| Electric output | = | 25.6 MW |
| Installed cost | = | $12.82 MM ($500/kW) |
| Capacity factor | = | 90% |
| Electric rate | = | $0.04/kWh |
| Fuel cost | = | $4.00/MMBtu |
| Steam price | = | $4.44/1000 lbs (83% of fuel cost) |
| General inflation rate | = | 6% |
| Electricity escalation rate | = | 8% |
| Fuel escalation rate | = | 10% |
| Steam escalation rate | = | 10% |
| Debt: equity ratio | = | 50:50 |
| Interest rate | = | 12% |

Ten-year project for economic evaluation purposes
Electric rate is an average value for all power sold
Installed cost includes equipment cost, installation, and project fees
75% of investment receives 5-year accelerated depreciation and 10% ITC
25% of investment receives 15-year depreciation and no ITC
Other costs include state and local taxes, insurance, and O&M cost
150# saturated steam valued at 1070 Btu/lb
Electricity valued at 3413 Btu/kWh

FIGURE 3-1. Comparison of Energy Utilization in a
Cogeneration System and a Conventional Powerplant.[1]

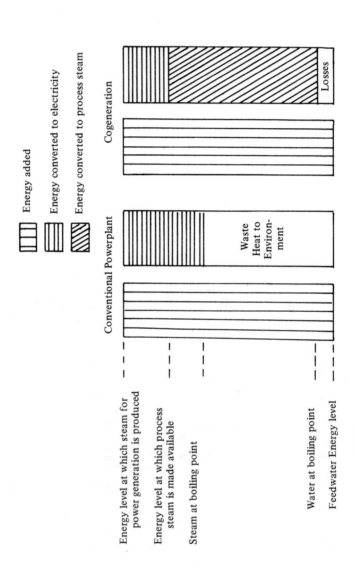

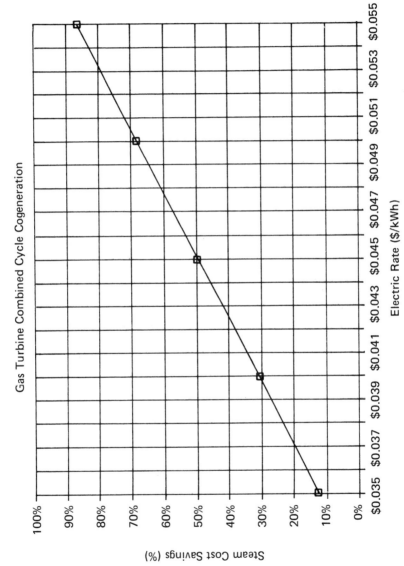

**FIGURE 3-2. Steam Cost Savings vs. Electric rate.**

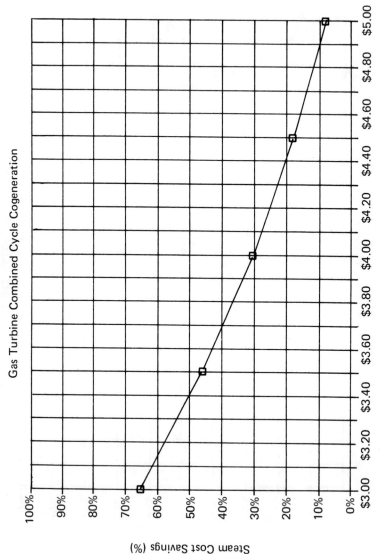

FIGURE 3-3. Steam Savings vs. Fuel Cost

Gas Turbine Combined Cycle Cogeneration

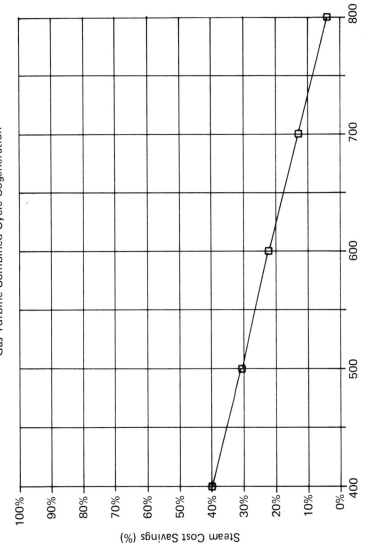

**FIGURE 3-4. Steam Savings vs. Capital Cost**

Gas Turbine Combined Cycle Cogeneration

# CHAPTER 4

# Optimizing Cogeneration in Industrial Plants for Energy Efficient Operation

*A.C. Sommer, M.A. Keyes, A. Kaya*

Industrial energy management is a complex problem. Although the problem may be handled by using the theory of large scale systems, it is more practical and economically feasible to design systems by empirically observing operational constraints and using sound engineering judgments.

Plant reliability, availability, maneuverability (responsiveness) and security considerations dictate that a system with distributed architecture and distributed computation be used. It is for this reason that currently available energy management systems of modern design have been developed accordingly.

## Description of Plant Equipment

The scope of energy management has been surveyed and summarized by Messrs. Kaya and Keyes, 1983. The energy flow for a pulp and paper plant is considered by three subsystems of Figure 4-1 (Kaya, 1978), namely:

1. Energy generation: boilers (including black liquor recovery), electric power generation.

2. Energy transmission: distribution equipment such as PRV; pumps, fans, and heaters of energy generation which are driven by steam or electricity as necessary; other auxiliaries for energy recovery, etc.

3. Energy consumption: process area primarily for production (including by-products of energy sources such as black liquor and bark).

There is strong interaction between these subsystems, and a multi-level approach is prudent. Few authors have identified their work in terms of multi-level control and optimization (Kaya, 1978; Womack, 1979; Balchen, 1979; Kociuba and Postingl, 1977; Aarnio, et at., 1980).

## Description of Multi-Levels

The levels are described as follows (Kaya and Keyes, 1983):

*On-line dedicated controls*: These controls perform specific functions and are designed to be responsive to process transients or disturbances. Simply, they control designated variables at the desired set point (set manually or automatically).

*Supervisory and optimization*: These functions perform optimizing control to help coordinate the operating levels (e.g., set points) of lower level controls so that the lower-level controls as a group provide the most efficient overall operation. The general idea here is to determine the value of set points of individual units so that the best performance of the subsystem operation is achieved while the demand is satisfied. A critical evaluation of such multi-level (decentralized) controls has been made by Javdan and Richards (1977). Optimization is performed over the total boiler, turbine and process system which is shown in Figure 4-1.

*Energy coordination and planning*: Higher level functions such as coordination between optimizing functions, future planning of production levels under postulated alternative energy availability and cost scenarios are performed at this level. The output of this level provides constraints and demand levels which must be satisfied by the lower level optimizing functions.

Kociuba and Ponstingl (1977) for example presented a 3-level energy management functional concept as control, supervisory, and planning. Similar to this, a hierarchy of responsibility and activity was proposed by Aarnio, et al., (1980).

## Description of Energy Coordination

The coordination level is the main topic of this chapter. This level deals with the optimum coordination of various power generation and purchase means to minimize the energy cost for a given total energy demand of a plant (Keyes and Kaya, 1983).

The structure of energy coordination is described in Figure 4-2. Optimum load allocation of boilers is assumed to meet a specified steam demand. Furthermore, the optimum extraction allocation of turbines is assumed to meet a specified process steam and/or in-plant electric generation demand, and tie-line control is assumed to be optimum when the demand charge is minimized. Also the operation of dual drives is optimized for such major equipment as pumps, fans and compressors. The energy coordination problem can be defined as follows:

Given:

1. Steam cost, $/hr, versus steam demand, lb/hr, function of the boiler system; and,

2. In-plant electric generation kW, versus steam flows, lb/hr, for the turbine system; and,

3. Electric purchase and sale formula; and,

4. Power requirement versus load (output) of different equipment such as dual drives; and,

5. Process steam and electric load demand of the plant.

Determine:

1. The demand level of steam for the boiler system,

2. The demand level of in-plant electric generation of the turbine system,

3. The demand level of purchased electric power; and,

4. The drive means (steam or electric) for auxiliaries.

so that the energy cost per unit product or energy cost for a given rate of production is minimized.

Clearly, this problem assumes that the given functional relations of subsystems are the results of similar optimum load allocations at the next lower hierarchical level. A higher level optimization, or "coordination" as referred to herein, will allocate the demand levels for each lower level to achieve a global, minimized solution for energy cost.

The methods and approches for overall energy management are discussed by Kaya and Keyes (1983).

## DEVELOPMENT OF
## THE ENERGY COORDINATION PROBLEM

The power plant and process together is treated at this higher level. Each subsystem is considered as an aggregated unit which performs according to a functional relation. The physical interactions between these subsystems should be taken into account for higher level coordination. A global description of the plant without the lower level details is described in Figure 4-3.

### Physical Relations of Subsystems

Each subsystem has a cost-based relation that is a result of optimization at lower levels. Referring to Figure 4-3, these relations are briefly described in the following.

*Boiler system.* The problem is defined as minimizing the fuel cost Cf, $/hr, by satisfying the demand $x_s$, lb/hr. So the functional relation is

$$C_f = f_f (x_s) \qquad (1)$$

Several authors have reported on boiler optimization (Cho, 1978; Leffler and Shizemura, 1978; Kaya and Keyes, 1983).

*Turbine system.* Turbine system optimization approaches are slightly more complicated than boiler systems since the energy from each of the several stages of extraction can be utilized in several ways. Futhermore, the turbines may not all be fed from the same pressure header.

The generating turbines are usually used for low cost electric generation while the steam demand of the plant is satisfied. At the same time, the turbines may be utilized for electric generation with no regard to steam demand, during the onpeak period by increasing condensing flows. These alternative uses must be simultaneously met by the turbine system.

Turbine optimization problems may be outlined as follows:

1. Electric demand-based optimization: For a given kilowatt demand of in-plant electric generation, the utility value of steam energy flow, $/hr, used due to the electric generation is minimized while the electric demand is satisfied.

2. Steam demand-based optimization: For given steam flows (demands) of pressure heads, lb/hr, the electric generation is maximized while the steam demand is satisfied.

As a result of these combined optimizations, a relation for the turbine system exists:

$$W_{15} = f_{15}(x_2, x_3, x_4, x_5) \tag{2}$$

where

$f_{15}$ = a function

$W_{15}$ = electric power generated, kW

$x_2, x_3, x_4, x_5$ = flows from turbine system, lb/hr

The individual turbine flows are of no concern as they are determined by the combined optimization.

Hanson (1978), Hunt (1979), Aarnio (1979) and other workers identified the turbine optimization and reported the results of their work. However, what should be emphasized is that the problem of turbine optimization must be posed in terms of the cost (utility) values of steam flows (Kaya) and Keyes, 1983).


## Dual Drives and Process

There is often operating flexibility incorporated into drives and process units at the time of plant design. Pumps, compressors, fans, etc. may be powered by electric motors or steam turbines. A single pump can sometimes be driven by a turbine or motor selectively. There may be several parallel pumps or compressors with different driving means. The process may, in addition, use different energy sources. There is, however, a limited flexibility in what sort of power is utilized. The following relations between the loads (output) of dual process and drives and the power inputs (steam and/or electric) exist.

$$x_{10} = f_{10}(P_{10}) \tag{3}$$
$$W_{17} = f_{17}(P_{17}) \tag{4}$$

where

f = a function

x = steam input flow, lbs/hr

W = electric power input (real) kW

p = load of drives and process

Note that the sum of p values are equal to the load of the dual system.

Electric Purchase Rate Schedule

There are two variables from which the rate schedule is developed. The schedule is described as,

$$C_e = f_e (W_e, VAR_e) \tag{5}$$

where

$f_e$ = a function

$C_e$ = bill for the period, $

$W_e$ = active (real) demand, kW

$VAR_e$ = reactive demand, kvar

The function $f_e$ can be very complicated. The values of $W_e$ and $VAR_e$ versus time should be known to determine the bill for the period.


## MULTILEVEL COORDINATION PROBLEM

Referring to the multilevel structure in Figure 4-2 and the plant structure in Figure 4-3, the energy coordination problem is presented as follows:

Statement of Coordination Problem:

$$\min C = C_f + C_e$$
$$= f_f (x_s) + f_e (W_e, VAR_e) \tag{6}$$

(where, $C_f$ and $C_e$ are defined by equations (1) and (5), subject to the constraints as follows. In these relations, x = steam flow, lbs/hr, and W = real electric power, kW.

Generation - flow relations of turbine system

$$W_{15} = f_{15} (x_2, x_3, x_4, x_5) \tag{7}$$

Mass balance of turbine system

$$x_s = x_2 + x_3 + x_4 + x_5 \tag{8}$$

Real power balance of tie line

$$W_e + W_{15} = W_{17} + W_{18} \tag{9}$$

Load demand relation for dual process and drives

$$P_{10} + P_{17} = P_a \tag{10}$$

Power and load relations of dual process and drives

$$x_{10} = f_{10}\ (P_{10})\ (a)$$
$$W_{17} = f_{17}\ (P17)\ (b) \tag{11}$$

Mass balance of dual process and drives

$$x_{10} = x_{11} + x_{12} \tag{12}$$

Mass balance of process

$$x_2 + x_7 + x_8 = x_{13} \tag{13}$$

Mass balance of steam system

$$x_{12} + x_{13} + x_5 = x_s \tag{14}$$

Mass balance of steam flow conduits

$$x_3 = x_7 + x_{10}\ (a)$$
$$x_8 = x_4 + x_{11}\ (b) \tag{15}$$

Some of the x and W values are given which are dictated by the operating loads of the plant. These given values are,

$$P_a, x_2, x_7, x_8, W_{18} \tag{16}$$

The variables to be determined are as follows,

$$x_s, x_3, x_4, x_5, W_{17}, P_{10}, W_{15}$$
$$x_{10}, x_{11}, x_{12}, W_e, W_{15} \tag{17}$$

In addition, the limits of the variables on mass flows and power generation exist as follows:

Limits on flows as necessary,

$$x_{min} \leqslant x \leqslant x_{max} \tag{18}$$

Limits for in-plant electric power generations,

$$W_{15(min)} \leqslant W_{15} \leqslant W_{15(max)} \tag{19}$$

These limits are known by the structures of subsystems and available for coordination problem.

In this optimization problem the real power is coordinated while the purchased reactive power, $VAR_e$, is kept at a target value.[1] Then, for this target value, the reactive power of the plant must be regulated by in-plant VAR generation, capacitors, and by load shedding in the case of emergencies. The line control and VAR control can provide the desired results.

## SOLUTION OF COORDINATION PROBLEM

The problem has in the past been solved by nonlinear programming and implemented in large computer systems. Here, the problem is solved by a powerful distributed computational system applied to the task of energy management. Because of this structure, the implementation must be kept in mind. The problem will first be simplified for computational tractability. Also, an incremental approach will be introduced so that it can be readily implemented by a modern distributed microprocessor system in real time.

The procedure of simplification is as follows:

1. Use equation (8) to eliminate $x_5$ and replace it by $x_2, x_3, x_4, x_5$.

2. Equations (13) and (14) are not needed since they will either be inactive constraints or be redundant.

After substituting the preceeding simplifications, the simplified problem is to minimize C in equation (6) subject to equations (7), (9), (10), (11), (12), (15), (16), (17). The variables to be solved for are $x_3, x_4, x_5, W_{17}, P_{10}, P_{17}, x_{10}, x_{11}, x_{12}, W_e, W_{15}$. Specifically, the optimization problem is written as,

---

[1] Reactive power can be included as a variable for the coordination problem. Practical considerations favor a target value which is adjustable.

$$\min_{x_3,\ x_4,\ x_5,\ W_e}$$

$$C = f^1 {}_f(x_3,\ x_4,\ x_5) + f_e\ (W_e,\ VAR_e)$$

(6)

Subject to:

$$W_{15} - f^1 {}_{15}(x_3,\ x_4,\ x_5) = \tag{7}$$

$$W_e + W_{15} - W_{17} = W_{18} \tag{9}$$

$$P_{10} + P_{17} = P_a \tag{10}$$

$$x_{10} - f_{10}(P_{10}) = 0 \tag{11a}$$

$$W_{17} - f_{17}\ (P_{17}) = 0 \tag{11b}$$

$$x_{10} - x_{11} - x_{12} = 0 \tag{12}$$

$$x_3 - x_{10} = x_7 \tag{15a}$$

$$x_8 - x_{11,} = x_4 \tag{15b}$$

From the nature of problem only active limits are stated as follows:

$$x \leqslant x_{max};\ \text{for}\ ;x_3,\ x_4,\ x_5 \tag{16}$$

$$W_{15} \leqslant W_{15}(max) \tag{17}$$

In addition, all variables must be non-negative. The right-hand side of the equality constraints are given along with the lower and upper bounds of variables which are on the left-hand side of the equality constraints. The objective function and some of the constraint relations are nonlinear, making it a nonlinear optimization problem.

There are numerous subroutines to solve this problem (Kuester and Mize, 1973). Those subroutines can be handled by the mini-computers widely used in the process industry. A simpler incremental gradient approach could be used iteratively to obtain an optimum solution (Kennedy, 1975).

Method of Incremental Optimization

Consider an optimization problem,

$$\text{Max } J = f(y) \tag{18}$$

Subject to constraints,

$$g_i(y) \geqslant b_i \quad i = 1,...,m \tag{19}$$

where, y is a variable vector

$$y = [y_1,....,y_n]^T. \tag{20}$$

Define error functions for each constraint as,

$$e_i(y) = 0 \qquad \text{if } g_i(y) \geqslant b_i \tag{21}$$
$$= g_i(y) - b_i \text{ otherwise}$$

Consider an augmented function F as,

$$F(y) = \sum_{i=1}^{m} p_i[e_i(y)]^2 + kf(y) \tag{22}$$

where, $p_i$ and k are positive constraints. Note that F is an unconstrained function including the constraints as penalties if equation (19) is not satisfied. The extremal values of F correspond to the necessary condition of an optimum when the constraints are satisfied. The solution for the gradient relations, gives a set of y values for the optimum solution.

$$\frac{\partial F}{\partial y_j} = 0 = \sum_{i=1}^{m} p_i \frac{\partial e_i}{\partial y_j} e_i + k \frac{\partial f}{\partial y_j} \tag{23}$$

Also, the steady state values, $(dy_j/dt) = 0$, of the differential equation,

$$\frac{dy_j}{dt} = - \sum_{i=1}^{m} p_i \frac{\partial e_i}{\partial y_j} e_i + k \frac{\partial f}{\partial y_j} \tag{24}$$

provide a set of y values satisfying equation (23). Note that the negative sign in front of the summation assures a stable trajectory approaching the steady state. Note that the sign makes no difference since $e_i = 0$ must be satisfied. Equation (24) can be treated by numerical methods until a steady state is reached.

However, a unique approach by using microprocessors with simple function blocks, and without any computer programming, is utilized for the solution. The method is demonstrated by a simulation problem. Figure 4-4 describes a flow chart of the proposed method. The P and J blocks of Figure 4-4 will be worked out for a simulation example by using microprocessor function blocks.

## SIMULATION PROBLEM

A numerical example of the optimization problem presented by equations (6) - (17) is treated in Appendix A. Appendix A gives the details of the solution of the proposed method by using the microprocessor function blocks.

### Results

The initial values were assigned according to a logical operating condition. The optimization procedure brought the operation into an optimum with a lower value of objective function, as shown in Table 4-1. Initial values approached to optimum values within a short time and with a smooth exponential delay.

The only noticeable optimization result is that the purchased electricity was cut back and in-plant electric generation was increased. However, any arbitrary operating conditions approach the same optimum result since the stability of solution trajectories are assured and finite values are reached as the optimum solution.

|  | Initial | Optimum |
|---|---|---|
| Objective fn | 931 | 892 |
| $x_3$ | 60 | 59.8 |
| $x_4$ | 240 | 244 |
| $x_5$ | 6 | 17.2 |
| $W_e$ | 4 | 0 |
| $x_{10}$ | 10 | 9.81 |
| $W_{15}$ | 38 | 42.3 |
| $W_{17}$ | 8 | 8.36 |
| $P_{10}$ | .96 | .939 |
| $P_{17}$ | 5.6 | 5.62 |
| $x_{11}$ | 10 | 5.47 |
| $x_{12}$ | 0 | 4.33 |

TABLE 4-1. Optimum Solution of Coordination Problem.

## CONCLUSIONS

It has been demonstrated that a complex energy management problem can be decomposed (distributed) among many levels of the control hierarchy for computational ease with major implementation advantages. It is clear that a distributed microprocessor system is best suited for the distributed computational scheme presented. A good example demonstrating the validity of this approach is the example treated by Kaya and Keyes (1983) for a turbine system optimization in which a full linear programming and a simple incremental cost optimization approach gave the same results.

Here, a better implementation is to use a unique means of solving the nonlinear programming problem along with an incremental optimization approach. A low-cost distributed microprocessor system without high level programming is used to implement this method. The advantages of a distributed microprocessor system have been stated previously.

## REFERENCES

Aarnio, D.E., Tarvaninen, H.J. and Tinnis, V., (1980). "An Industrial Energy Management System" TAPPI, 73.

Balchen, J.G., (1979). "The Applicability of Modern Control Theory Related to Dynamic Optimization in Industry Today." *Control Engineering*. On-Line Optimization Techniques in Industrial Control, pp 51-62.

Cho, C.H. (1978). "Optimum Boiler Load Allocations, Instrumentation," *P&P Industry, Vol. 17*. Instrument Society of America, pp 39-44.

Javdan, M.R. and Richards, R.J., "Decentralized Control System Theory, a Critical Evaluation," *Intl. J. Control*, p. 129.

Kaya, A. (1978) "Industrial Energy Control: The Computer Technology in Pulp, Paper, and Allied Industries," *IFAC Journal of Automatica*, Vol. 19, No. 2, pp 111-130.

Kennedy, J.P., "A Simple Steady State Multivariable Control Algorithm with Examples of Gasoline Blending and Ammonia Reform Furnace," Summer Simulation Conference, July 21-24, 1975, San Francisco, CA, USA.

Keyes, M.A. and Kaya, A., "Plant Energy Management by Multi-Level Coordination," 5th Intl. IFAC-PRP/5 Conference, October, 1983, Antwerp, Belgium.

Kociuba, T. and Postingl, J. (1977), "Energy Management Systems for Industry," *Control Engineering Energy Conservation*, pp 43-56.

Kuester, J.L. and Mize, J.H. (1973), "Optimization Techniques with Fortran," Chapter 10, Box Complex Algorithm, McGraw-Hill.

Leffler, N. (1978), "Optimization of Coengineering," TAPPI Engineering Conference, pp 641-645.

Womack, B.F. (1978). "A Review of the Status of Optimization Theory," *Control Engineering*, On-Line Optimization Techniques in Industrial Control, pp 1-15.

## APPENDIX - A

The following substitutions of the variables and the modifications of constraints are made to bring the problem into the form used in the incremental optimization: $y_1 = x_3$; $y_2 = x_4$; $y_3 = x_5$; $y_4 = W_e$; $y_5 = x_{10}$; $y_6 = W_{15}$; $y_7 = W_{17}$; $y_8 = P_{10}$; $y_9 = P_{17}$; $y_{10} = x_{11}$; $y_{11} = x_{12}$.

Since min [C] corresponds to max [−C] the problem is,

$$\max [-f(y)] = -k[.0006 (y_1 + Y_2 + y_3)^2 +$$
$$2.6(y_1 + y_2 + y_3) - 12 + 6.74 + 21.1 \, y_4] \tag{A1}$$

Constraints are,

$$g_1(y) = y_6 - .125y_1 - .25y_2 - .222y_3 = -30.4 = b_1 \tag{A2}$$

$$g_2(y) = y_4 + y_6 - y_7 = 34 = b_2 \tag{A3}$$

$$g_3(y) = y_2 + y_9 = 6.56 = b_3 \tag{A4}$$

$$g_4(y) = y_5 - 9.55y_8^2 - 4 \, y_6 = 1 = b_4 \tag{A5}$$

$$g_5(y) = y_7 - .027 \, y_9^2 - .88 \, y_9 = 2.6 = b_5 \tag{A6}$$

$$g_6(y) = y_5 - y_{10} - y_{11} = 0 = b_6 \tag{A7}$$

$$g_7(y) = y_1 - y_5 = 50 = b_7 \tag{A8}$$

$$g_8(y) = y_2 + y_{10} = 250 = b_8 \tag{A9}$$

$$g_9(y) = -y_1 \geqslant -260 = b_9 \tag{A10}$$

$$g_{10}(y) = -y_2 \geqslant -280 = b_{10} \tag{A11}$$

$$g_{11}(y) = -y_3 \geqslant -80 = b_{11} \tag{A12}$$

$$g_{12}(y) - y_6 \geqslant -44.6 = b_{12} \tag{A13}$$

$$g_i + 12(y) = y_1 \geqslant 0 = b_i + 12 \tag{A14}$$

The error functions are, (for $p_i = 1$)

$$e_1 (y) = g_1 (y) - b_1 \text{ if } g_1 (y) \neq B_1 = 0 \text{ otherwise} \tag{A15}$$

The same holds for $e_2$ through $e_8$. For the rest.

$$e_9 (y) = g_9 (y) - b_9 \text{ if } g_9 > b_9 = 0 \text{ otherwise} \tag{A16}$$

The same holds for $e_{10}$ through $e_{23}$.

A general block diagram is given in Figure 4-4. The values of

$$P_i = 1; k = .1 \tag{A17}$$

will be used. Only the case for $j = 1$ will be shown to cover all variety of function blocks. Referring to Figure 4-4, only non-zero values of $(\partial e_i / \partial y_j)$ and $(\partial C / \partial y_j)$ will be needed.

**Diagram for $y_1$** (Figure 4-5)

$$(\partial e_1 / \partial y_1) = .125; (\partial e_7 / \partial y_1) = 1.$$

$$(\partial e_g / \partial y_1) = -1.; (\partial e_{13} / \partial y_1) = 1$$

$$(\partial C / \partial y_1) = .0012 (y_1 + y_2 + y_3) + 5.2$$

The corresponding $e_1$, $e_7$, $e_9$, $e_{13}$ values are substituted from (A15) and (A16) to structure the microprocessor diagram. Note that $(\partial a_i / \partial y_1)$ values are constant as a special case. In general $y_j$ values are used to generate $\partial e_i / \partial y_j$.

The function blocks in Figure 4-5 are commercially available and represent "the state-of-the-art" of microprocessors.[1] The solution approached the following values at steady state which is optimum and all constraints were satisfied.

|  |  |  |
|---|---|---|
| C = 892 | $y_6 =$ | 42.3 |
| $y_1 =$ 59.8 | $y_7 =$ | 8.36 |
| $y_2 =$ 244 | $y_8 =$ | .939 |
| $y_2 =$ 17.2 | $y_9 =$ | 5.6 |
| $y_4 =$ 0 | $y_{10} =$ | 5.47 |
| $y_5 =$ 9.81 | $y_{11} =$ | 4.33 |

(A18)

[1] Application Manual of NETWORK 90 Microprocessor Systems, Bailey Controls Co.

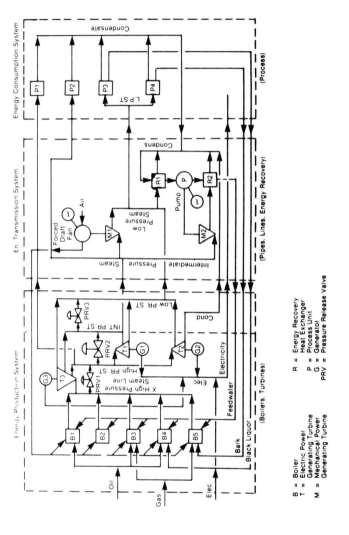

**FIGURE 4-1. Energy Management System Model for Paper and Pulp Plant.**

B = Boiler
T = Electric Power
   = Generating Turbine
M = Mechanical Power
   = Generating Turbine

R = Energy Recovery
   Heat Exchanger
P = Process Unit
G = Generator
PRV = Pressure Release Valve

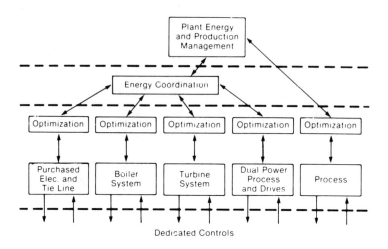

FIGURE 4-2.  Multilevel Structure of Energy Management.

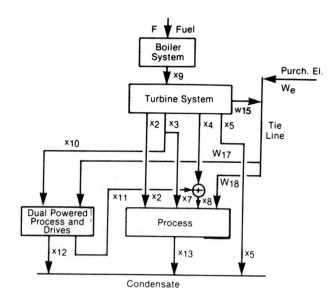

FIGURE 4-3.  Global Description of Plant.

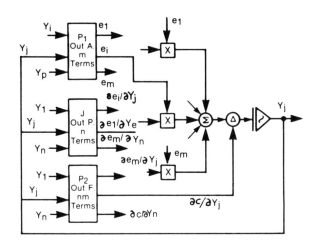

**FIGURE 4-4.** Flow Chart of Incremental Optimization.

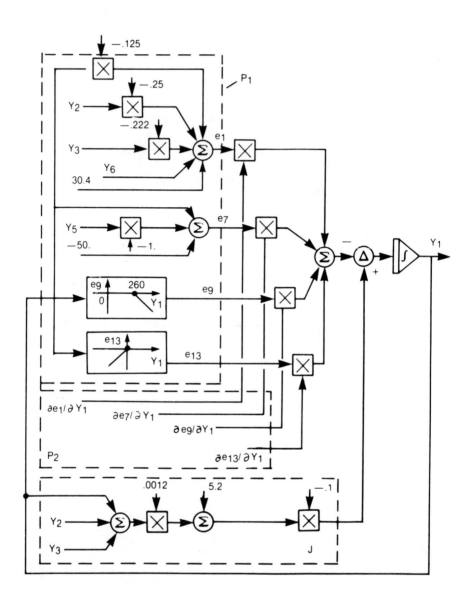

**FIGURE 4-5.** Microprocessor Diagram for $Y_j$.

# CHAPTER 5

# Upgrading Thermal Output by Mechanical Vapor Recompression

*F.E. Becker*

In cogeneration systems utilizing reciprocating engines, approximately 30 percent of the input fuel energy is dissipated to the engine cooling jacket and can only be directly recovered for process use in the form of hot water or low-pressure steam. Because the need for high-pressure steam is usually of greater importance, there is significant merit in being able to raise the pressure level of the steam produced in the cooling jacket from ebullient cooling to process level conditions.

The most practical approach to raising the pressure level is by mechanical vapor recompression. Low-pressure steam contains significant energy in the form of latent heat and requires relatively little additional energy input in the form of the compression work to raise its pressure to process level conditions. While it is recognized that many processes have a large need for hot water, this can usually be supplied from many other low-level heat sources presently not being recovered.

Also, since additional high-pressure steam can be produced directly in a waste heat boiler using the engine's high-temperature exhaust gas, the total process heat produced is of a single high level nature (high-pressure process steam) instead of varying proportions of high- and low-pressure steam or hot water.

This chapter describes the merits of mechanical vapor recompression for upgrading low-pressure cooling jacket steam and several approaches which can be utilized to integrate a mechanical vapor recompression system (MVRS) into the cogeneration plant.

## BASIC CONCEPT

Low-pressure steam has long been considered a low-level heat source and of little practical value. Energy engineers tend to "look" for places to use this steam, even though the same heat requirements can often be met from other low-level waste heat sources. However, no one likes to see steam plumes and thus this steam often ends up making hot water (of which there is usually already an excess) or is used for plant heating.

Low-pressure steam, however, contains significant energy as latent heat; in fact, in a typical direct-fired boiler, 80 percent of the fuel energy enters the steam as latent heat. The underlying problem with low-pressure steam is that its corresponding saturation temperature is often too low to transfer its latent heat to the desired process.

Adding thermal energy will simply raise its temperature to make superheated steam, and will not enable the recovery of the latent heat which will still be constrained by the saturation pressure and associated temperature. In order to recover the latent heat at a given temperature, the steam pressure must be raised to the desired associated saturation temperature. This is most easily accomplished by mechanically recompressing the steam.

The specific amount of work that must be done on the steam to raise its pressure to a given level can be determined with reasonable accuracy, assuming an ideal gas, by the following relation:

$$W = \frac{C_p T_1}{n_m n_s} \left[ 1 - \left( \frac{P_2}{P_1} \right)^{\frac{k-1}{k}} \right]$$

where:

$C_p$ = specific heat of steam

$k$ = specific heat ratio of steam ($C_p/C_v$)

$P_1$ = compressor inlet pressure

$P_2$ = compressor outlet pressure

$T_1$ = compressor inlet steam temperature

$W$ = specific power requirement

$n_m$ = compressor mechanical efficiency

$n_s$ = compressor isentropic efficiency

Figure 5-1 is a plot of the typical compressor power requirement for recompressing 15 psig cooling jacket steam to various process pressure levels.

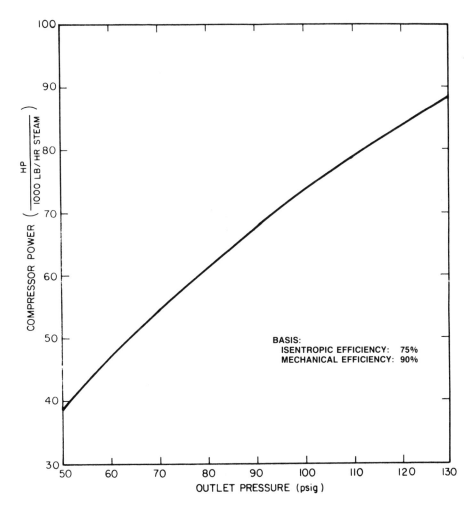

**FIGURE 5-1. Compressor Power Requirement —
Inlet Pressure 15 psig.**

The power requirement is a strong function of the pressure ratio, and thus to minimize the work input, the pressure should be increased only to the level required by the process, and not necessarily the main steam header pressure. While the power requirement may seem large, with respect to the latent heat already in the steam, it is relatively small.

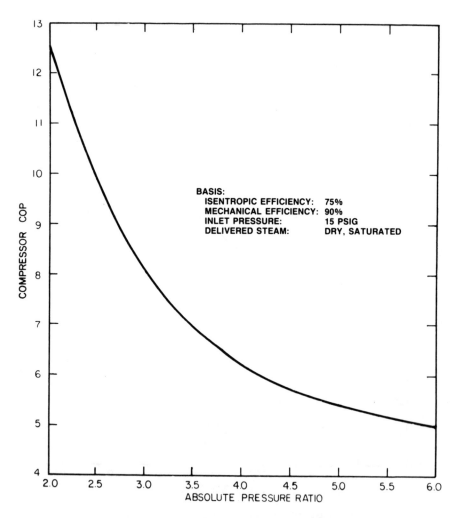

FIGURE 5-2. Compressor Coefficient of Performance at
Various Pressure Ratios.

For example, 15 psig steam contains 954 Btu/lb of latent heat. Mechanically recompressing this steam to 125 psig, or a pressure ratio of 4.67, requires an additional work input of only 190 Btu/lb or 20 percent. Of course, this energy input is not lost because it is added to the steam.

The ratio of the energy output of the steam to the work input to the compressor is often referred to as the compressor Coefficient of Performance or COP. Figure 5-2 shows the compressor COP for various pressure ratios. While the curve is drawn for 15 psig inlet steam, it is also fairly representative of other inlet pressures.

Again, it is important to note the strong relationship between the pressure ratio and the compressor COP and, for maximum COP, the pressure should be raised only to the level actually required by the process. However, even at pressure ratios up to 6, respectable COP's of 5 can be achieved.


## DESIGN APPROACH

Two basic design approaches may be utilized for integrating mechanical vapor recompression into a reciprocating engine-driven cogeneration system. In both cases, a positive displacement rotary screw compressor is recommended for recompressing the steam. The details for this selection are discussed later.

The first, shown schematically in Figure 5-3, uses the same engine to drive both the generator and compressor. Therefore, the engine would have to be oversized to take care of the added compressor load. For small, base loaded systems, where there is only one engine-generator set, this is a good design approach. A feature of this setup is that if high-pressure steam is not always needed, and there is an increased demand for electric power, the compressor could be unloaded or decoupled and advantage taken of the additional engine power available.

A complication with this scheme, however, is that under conditions of varying electric power demand, the flow rate of low-pressure steam from the engine cooling jacket would change. This would require modulating the compressor throughput without changing the engine speed since this would alter the generator frequency. Modulation could be accomplished by: (1) recirculating using a por-

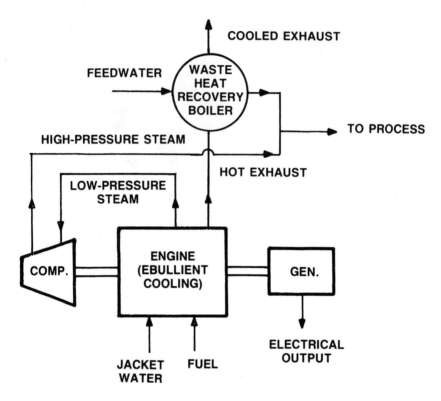

FIGURE 5-3. Individual Engine-Generator Set With MVRS.

tion of the recompressed steam back to the inlet, (2) use of an inlet throttle valve, (3) speed control using a variable speed fluid coupling, or (4) the use of a special "slide valve" which changes the compressor capacity. The first two options would not save any power at the reduced steam flow rate, but would satisfy the system operational requirements.

The second scheme, shown in Figure 5-4, would utilize one or more compressors, each driven by a dedicated driver. This choice is recommended for multiple engine-generator sets.

The vapor recompression system, shown in Figure 5-5, consists of a rotary screw compressor, engine prime mover, speed increaser (gearbox), waste heat recovery boiler, and control system. The low-pressure steam from several engine-generator sets, as well as the compressor driver, would be consolidated and piped over to a single compressor. In this fashion, the compressor could be better sized to take advantage of the lower specific cost of larger machines.

Also, by not directly linking the compressor to the generator, there is greater overall operating flexibility. Compressor throughput could be varied by any of the methods previously mentioned without affecting the operation of the engine-generator. However, since speed regulation is the most efficient method for modulating compressor throughput, the built-in variable speed capability of the engine would provide an ideal match-up.

For both schemes, high-pressure steam would be produced in exhaust waste heat recovery boilers. These could be individually matched to each engine or the exhaust gases consolidated and sent to fewer, but larger, boilers.

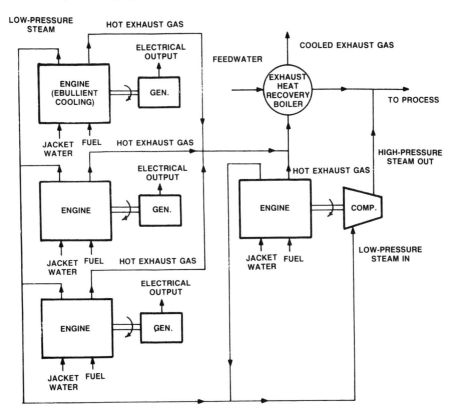

**FIGURE 5-4. Multiple Engine-Generator Sets With Single MVRS.**

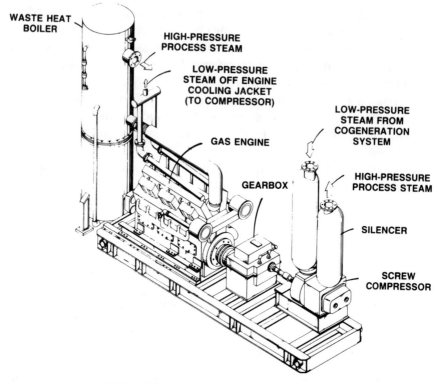

**FIGURE 5-5.** Vapor Recompression System.

## COMPRESSOR SELECTION

In the selection of the compressor, there are a number of important considerations:

1. Mechanical reliability
2. Efficiency
3. Capacity
4. Turndown capability
5. Cost

First, the compressor must be specifically designed for recompressing steam. Although there are numerous types of compressors, there are usually some unique operational limitations which must be considered when recompressing steam. For example, the high impeller tip speeds of centrifugal compressors make them highly susceptible to erosion from entrained water droplets. This erosion can reduce the efficiency as well as cause dynamic instability from rotor imbalance and mechanical failure.

Next, it is of obvious importance to minimize the power input to the compressor; therefore, the compressor should have high mechanicle (>90 percent) and isentropic (>75 percent) efficiencies. Except for mechanical and jacket heat losses, the power input to the compressor ends up in the steam in the form of sensible heating and pressure rise. The greater the compressor efficiency, the more this power ends up as pressure rise of the steam—the ultimate goal—and not as sensible heating.

Next, the capacity of the compressor should be well matched to the cogeneration system. Single machines that do not have adequate flow capability would require that several be operated in parallel, increasing complexity of the installation. Too large a compressor generally results in reduced performance at off-design conditions, which leads to the question of turndown capability of the compressor. In cogeneration installations with varying electric power demand, the low-pressure steam flow would vary, and the compressor must be capable of responding to this situation. A two-to-one turndown capability would normally prove adequate.

Lastly, the cost of the compressor must be considered in terms of the previously mentioned factors. If the compressor is too expensive, it may not give an adequate return on investment to justify its use.

Compressors can generally be classified into two types, aerodynamic and positive displacement, as shown in Figure 5-6. Among the various types of compressors, centrifugal, reciprocating, Roots, and screw have been used for steam recompression.

### Centrifugal Compressors

Centrifugal compressors are the most widely used for steam recompression. These machines achieve compression based upon aerodynamic operating principles. To achieve the desired pressure rise, the machines run at relatively high impeller tip velocities, up to 1200 ft/sec. The machines are very sensitive to water droplets that can cause blade erosion, rotor imbalance, and potential failure. To avoid

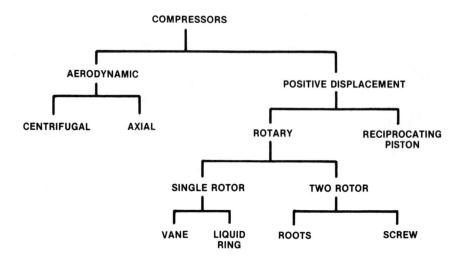

**FIGURE 5-6.** Classification of Mechanical Compressor Types.

this problem about 10 percent of the outlet superheated vapor is recirculated back to the inlet to dry the incoming steam.

Pressure ratios for centrifugal compressors are generally limited to two in a single stage. To achieve the pressure ratios typically required for cogeneration applications multistage machines would be needed. The machines are essentially constant pressure devices, and the power consumption is almost directly proportional to the volume delivered. Centrifugal compressors come in capacity ranges from 3,000 to 180,000 cfm.

Capacity control is most efficiently achieved by speed variation or the use of inlet guide vanes. With speed control, there is a minimum speed below which operation becomes unstable, a condition known as surge.

The surge limit is set by the impeller discharge angle and is generally around 50 percent of the flow capacity at the maximum efficiency point. In addition to speed control, the use of inlet vanes is common practice. The guide vanes act to provide prerotation ahead of the impeller to reduce entrance losses and as a throttle to reduce the flow rate by reducing the vapor density.

## Reciprocating Compressors

Reciprocating compressors have been available for many years in air and gas service and have reached a high degree of development. More recently, they have been applied to steam. They come in size ranges from 15 to 15,000 cfm and are furnished either as single stage or multistage.

The number of stages is determined by the compression ratio. The compression ratio per stage is generally limited to four, although small-size units are furnished with a compression ratio of up to 8 or higher.

A major operating problem with reciprocating compressors is the need to keep the steam dry. Any liquid at the inlet to the compressor is very detrimental to the mechanical reliability because it causes excessive wear of valves, pistons, and rings. In most applications, an effort is made to superheat the incoming vapor and/or separate out the liquid droplets. Although the machines are made to be oil free by using carbon/Teflon rings, there is still a small amount of oil carry-over which must be considered.

## Roots Blowers

Roots blowers are positive displacement machines. These machines are constant volume devices. The discharge pressure is determined by the resistance of the system. Blowers are generally used at low pressures, and low-pressure rises up to 18 psi due to excessive leakage, "slip," and rotor defection. Roots blowers can handle wet inlet steam.

Capacity ranges are from 200 to 25,000 cfm. Flow delivery is varied by changing the speed or bypassing some of the machine capacity. Bypassing offers no energy savings. The major limiting factor for Roots blowers is the limited pressure differential. To achieve the desired pressure boost for most applications, several machines operating in series would be required.

## Helical Rotary Screw Compressor

The helical rotary screw compressor is considered the best suited machine for steam recompression in cogeneration systems. It is a positive displacement machine consisting of two mating helically grooved rotors, one male and the other female, operating in a stationary housing with suitable inlet and outlet ports. No inlet or discharge valves are required.

Figure 5-7 is a photograph of a typical screw compressor with the upper half of the casing removed. The two rotors are supported at both ends of the casing by sleeve bearings which carry the radial loads. The rotors are prevented from contacting each other by a set of synchronizing gears. The gears have to transmit only about 10 percent of the driving power. The casing is water jacketed to assure dimensional stability.

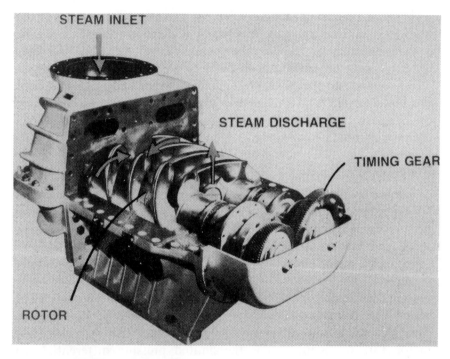

**FIGURE 5-7. Screw Compressor With Upper Casing Removed.**

In operation, inlet gas is pulled into a void created by a pair of spiral rotors, as shown in Figure 5-8. As rotation occurs, the incoming gas is cut off from the inlet and compressed by the meshing rotors as it is moved along the axis of the machine. At some point, depending on the compression ratio built into the machine, discharge ports are uncovered and the compressed gas is discharged. As with Roots blowers, the discharge pressure is a function of the resistance on the discharge side.

This type of machine is built in a wide range of sizes and is characterized by good volumetric and adiabatic efficiencies over a range of 50 to 100 percent of maximum capacity. Flow capacities up to 22,500 acfm are available in a single unit.

FIGURE 5-8. Screw Compressor Operation.

Figure 5-9 illustrates a typical performance map of a screw compressor. Flow and horsepower are proportional to speed, and speed variation is the most efficient method of capacity control. In addition to speed variation, machines larger than 5000 acfm have been built using a "slide valve" to control capacity by changing the effective rotor length or displacement. Three-to-one turndown can be achieved with a slide valve.

An internal compression ratio is built into the machine, and operation at an external pressure ratio other than the built-in compression ratio results in some loss in efficiency. Screw compressors are available with built-in compression ratios up to seven for single-stage units.

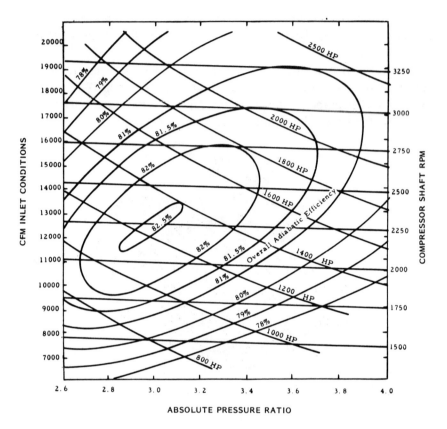

INLET PRESSURE = 14.7 PSIA
BUILT-IN PRESSURE RATIO = 3.0

**FIGURE 5-9. Typical Screw Compressor Performance Map
for Steam Recompression.**

The screw compressor has a number of unique characteristics which make its application well suited for steam recompression. Because of the relatively low velocities and smooth fluid flow path along the rotors, the machine is capable of compressing wet steam without damage. In addition, at high-pressure ratios, intercooling is not necessary because the work which normally goes into producing sensible heating during the compression process simply causes additional fluid to evaporate. This makes the process more nearly isothermal and, in doing so, minimizes the work of compression.

Several different thermodynamic paths may be followed during the compression process. These paths, shown in Figure 5-10, are: (1) compression of a two-phase wet mixture of appropriate quality to final conditions, (2) compression of dry saturated steam to final pressure, with final temperature attained by the addition of liquid to the superheated vapor, and (3) multistaged compression with liquid addition between stages.

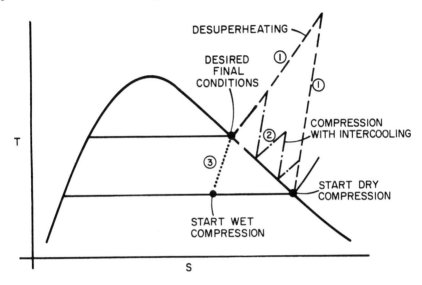

① DRY COMPRESSION WITH DESUPERHEATING - LEAST EFFICIENT

② DRY COMPRESSION WITH INTERCOOLING - MORE EFFICIENT

③ WET COMPRESSION WITH SCREW COMPRESSOR - MOST EFFICIENT AND COST EFFECTIVE

**FIGURE 5-10. Possible Steam Compression Paths.**

The screw compressor, because it is able to compress wet steam, has the potential of following the first path, which requires the least work of the three potential paths because the fluid remains homogeneous and in thermodynamic equilibrium during compression. If the steam were to first be dried, as would be required for centrifugal machines to avoid erosion damage, the second or third paths would be followed, both of which are less efficient than the potential path followed by the screw compressor. The theoretical increase in performance of a compressor operating under these alternate conditions is shown in Figure 5-11. At pressure ratios around five, improvements in compressor performance of up to 27 percent are possible.

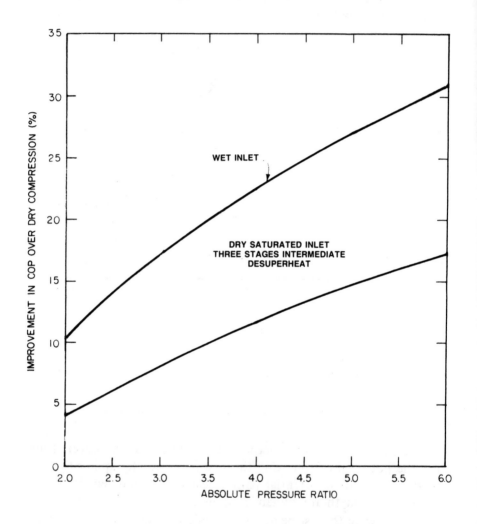

FIGURE 5-11. Increase in Compressor Performance Over
Dry Compression With Inlet and Interstage Water Injection.

The screw compressor also has another unique feature in comparison to reciprocating and centrifugal machines. Whereas contaminants in the steam can cause accelerated wear of the sliding parts of reciprocating machines and possible erosion in centrifugal machines, such a buildup of material in a screw machine actually improves the performance. This buildup will continue only to the extent allowed by the running clearances. Any excess gets carried out with the steam. The result is a higher delivered volume at the same power consumption.

## RECOMPRESSION ECONOMICS – CASE STUDIES

The economic justification for recompression of the cooling jacket steam is directly dependent on the value the user places on the low-pressure steam: the lower the value, the greater the justification for recompression. If there is a direct use for the low-pressure steam nearby and there are no alternative sources of heat that could also satisfy the process requirements, then a good plant energy balance has been achieved and there is no need for recompression.

In this instance, the user may consider the value of the low-pressure steam equal to that of the high-pressure boiler steam. However, sometimes the low-pressure steam is at one end of the plant and its potential use is a long distance away. In this case the value of the steam is lower because a low-pressure pipeline must now be built to make use of the steam. The most frequent case, however, is that there are many alternative sources of low-grade heat which could be used in place of the steam, and the potential use of the steam is a long distance away. Under this circumstance, the user would consider the low-pressure steam of almost no value.

For this last instance where the low-pressure steam is considered of no value, six case studies have been prepared, as shown in Table 5-1. These case studies represent electrical outputs of 1.5, 3.0, and 4.5 MW, and process steam pressures of 60 and 110 psig. Systems larger than 4.5 MW, e.g., 9 MW, would simply be considered multiple units.

These cases illustrate the impact that the size of the cogeneration system, the steam flow rate, and the pressure level to which the steam is being raised have on the system costs and payback periods.

## TABLE 5-1. Mechanical Vapor Recompression System Economics.

| | | CASE 1 | CASE 2 | CASE 3 | CASE 4 | CASE 5 | CASE 6 |
|---|---|---|---|---|---|---|---|
| **COGENERATION – MVRS OUTPUTS** | | | | | | | |
| Electrical Output | MW | 1.5 | 1.5 | 3.0 | 3.0 | 4.5 | 4.5 |
| Process Steam Pressure | psig | 110 | 60 | 110 | 60 | 110 | 60 |
| Steam Flow – Source Breakdown | | | | | | | |
| Cogeneration System – Cooling Jacket | lb/hr | 5,117 | 5,117 | 10,234 | 10,234 | 15,351 | 15,351 |
| – Exhaust | lb/hr | 2,815 | 2,815 | 5,630 | 5,630 | 8,445 | 8,445 |
| MVRS – Cooling Jacket | lb/hr | 1,400 | 801 | 2,800 | 1,602 | 4,200 | 2,403 |
| – Exhaust | lb/hr | 735 | 441 | 1,470 | 882 | 2,205 | 1,323 |
| – Injection Water | lb/hr | 886 | 473 | 1,772 | 946 | 2,658 | 1,419 |
| Total Steam Flow | lb/hr | 10,953 | 9,647 | 21,906 | 19,294 | 32,859 | 28,941 |
| **MVRS OPERATING SPECIFICATIONS** | | | | | | | |
| Inlet Pressure | psig | 15 | 15 | 15 | 15 | 15 | 15 |
| Outlet Pressure | psig | 110 | 60 | 110 | 60 | 110 | 60 |
| Pressure Ratio | | 4.2 | 2.5 | 4.2 | 2.5 | 4.2 | 2.5 |
| Inlet Steam Flow Rate[1] | lb/hr | 6,517 | 5,918 | 13,034 | 11,836 | 19,551 | 17,754 |
| Outlet Steam Flow Rate[2] | lb/hr | 7,403 | 6,391 | 14,806 | 12,782 | 22,209 | 19,173 |
| Compressor Power | hp | 525 | 315 | 1,050 | 630 | 1,575 | 945 |
| **MVRS STEAM COSTS** | | | | | | | |
| Annual Operating Time | hr | 8,400 | 8,400 | 8,400 | 8,400 | 8,400 | 8,400 |
| Gas Rate | $/MMBtu | 6.00 | 6.00 | 6.00 | 6.00 | 6.00 | 6.00 |
| Steam Flow[3] | lb/hr | 8,138 | 6,832 | 16,276 | 13,664 | 24,414 | 20,496 |
| Energy Input | MMBtu/hr | 4.14 | 2.48 | 8.28 | 4.96 | 12.42 | 7.44 |
| Annual Energy Cost | $ | 208,656 | 125,194 | 417,312 | 250,388 | 625,968 | 375,582 |
| Annual Maintenance Cost | $ | 26,460 | 15,876 | 52,920 | 31,752 | 79,380 | 47,628 |
| Total Steam Cost | $ | 235,116 | 141,070 | 470,232 | 282,140 | 705,348 | 423,210 |
| | $/1000 lb | 3.44 | 2.46 | 3.44 | 2.46 | 3.44 | 2.46 |

*(continued)*

TABLE 5-1. Mechanical Vapor Recompression System Economics *(continued)*.

| | | | | | | | |
|---|---|---|---|---|---|---|---|
| **CURRENT BOILER STEAM COST** | | | | | | | |
| Steam Flow(4) | lb/hr | 8,138 | 6,832 | 16,276 | 13,664 | 24,414 | 20,496 |
| Steam Cost | $/1000 lb | 7.75 | 7.75 | 7.75 | 7.75 | 7.75 | 7.75 |
| Total Steam Cost | $ | 529,784 | 444,763 | 1,059,568 | 889,526 | 1,589,352 | 1,334,289 |
| **SAVINGS** | | | | | | | |
| Annual | $ | 294,668 | 303,693 | 589,336 | 607,386 | 884,004 | 911,079 |
| | $/1000 lb | 4.31 | 5.29 | 4.31 | 5.29 | 4.31 | 5.29 |
| Percent | % | 55.6 | 68.3 | 55.6 | 68.3 | 55.6 | 68.3 |
| **MVRS INSTALLED COST** | | | | | | | |
| • Skid-Mounted Compressor, Engine Gearbox, Lube Console, Waste Heat Boiler, Controls and Instrumentation, Installation | $ | 600,000 | 525,000 | 865,000 | 715,000 | 1,055,000 | 835,000 |
| SIMPLE PAYBACK | yr | 2.03 | 1.72 | 1.47 | 1.18 | 1.19 | 0.92 |

NOTES

(1) Combined low-pressure cooling jacket steam from cogeneration and MVRS engines.

(2) Includes compressor desuperheating injection water.

(3) Total of compressor output plus direct high-pressure steam produced from MVRS engine exhaust.

(4) Equivalent steam output of MVRS. See Note (3).

*(end)*

These costs and payback periods represent only the recompression portion of the total cogeneration system. The cost and performance analyses are based upon generalized curve fits developed from manufacturers' data.

Three major factors are evident from these six examples. First, a significant fraction of the total process steam flow—17.8 percent for 60 psig steam and 27.6 percent for 110 psig steam—is contributed from the MVRS engine cooling jacket and exhaust heat, and compressor desuperheating injection water. This additional steam production is a direct result of incorporating mechanical vapor recompression into the cogeneration system.

Second, there is a significant reduction from two years to one year in the simple payback periods for increasingly larger systems. Thus, for multiple engine-generator sets, it makes more sense to consolidate the steam from several engines than to have a separate smaller unit for each.

Third, for lower recompressed steam pressure levels, the power input decreases significantly along with the payback period. As previously stated, it is best to recompress the steam only to the pressure level needed by the process. Of course, there are circumstances where the use is not close by and it becomes much simpler to raise the pressure to the common header level for wide-spread use.

The most significant aspect, however, is that recompression of the low-pressure steam off the cooling jacket is most economically attractive with simple payback periods of one to two years. This is generally in the same range as the cogeneration system by itself and, thus, vapor recompression does not tend to decrease the initial impetus for cogeneration, but adds to it.


## CONCLUSION

The use of mechanical vapor recompression to upgrade low-pressure cooling jacket steam offers a simple, viable, economically attractive approach for increasing the energy level of process heat from cogeneration systems. In addition, it offers a means for improving the energy efficiency of the entire plant. Once low-pressure steam is used for a low-grade application, the practicality of recovering even lower grade waste energy, which could also be used for the same application, becomes more remote. By mechanically recompressing low-pressure steam, a better match-up can be made between low-grade waste energy sources and process uses, thereby improving the plant energy efficiency from the aspects of both cogeneration and waste heat recovery.

# CHAPTER 6
# Modular Cogeneration for Commercial/Light Industrial Firms

*R. Sakhuja*

While increases in the cost of energy have improved the prospects for economically viable cogeneration, it was this factor plus other institutional changes that suddenly made small-scale cogeneration economical. Electric utilities are now required, under the Public Utilities Regulatory Policy Act (PURPA) of 1978, to permit qualified cogenerators to connect in parallel to their distribution system, to treat them without prejudice, and to pay "avoided costs" to them for power that they produce.

While in some regions the PURPA provisions are hotly contested or passively ignored, there are a number of major regions where electric utilities are friendly or neutral toward cogenerators. This is particularly true where electric power is in short supply and the cost of new central power stations is seen as excessive by the utility. State public utility commissions have an active role in the promulgation and enforcement of PURPA regulations, and have in some states led the fight for cogeneration.

As recently as the late 60s an attempt was made to market small cogeneration systems by the gas utilities. This effort was abandoned because of unsatisfactory experience. A study of this venture leads to some very interesting observations. These systems were installed in an era when electric utilities were unfriendly to cogeneration. As a result, the cogeneration equipment was essentially grid isolated. It had to have stand-alone capability and reliably meet all of a facility's electric load.

This required large capital expenditures for generating capacity sufficient to cover peak electric demand and occasional equipment failures or scheduled shutdowns. Utilization of both capital equipment and engine waste heat was, in general, poor. So too, were the system's economic returns.

Parallel operation of cogeneration systems with the electric utility's distribution system allowed under PURPA permits sizing of the cogeneration equipment relative to the site's thermal load thus ensuring better utilization of the available thermal output of the cogeneration system.

Another key reason for unsatisfactory performance of these systems was the system maintenance. This was, in part, due to lack of clear identity of the responsible organization for this task. In many instances, customers assumed the maintenance responsibility but were inadequately equipped to handle it. This makes a very strong case for single-source responsibility for maintenance. A cogenerator supplying hot water and electricity has to demonstrate to the user a trouble-free operation comparable to his current equipment; i.e., his water heater.

Most of the commercial and light industrial establishments who can benefit from this system may not have appropriate operating and maintenance personnel on the premises. A responsible manufacturer of such cogeneration equipment can only divest himself of the maintenance responsibility at a great risk of poor system operation and a potentially bad reputation for the entire industry.

## SYSTEM DESCRIPTION

The underlying design philosophy of a successful small cogeneration system has to be high reliability at low cost. Low cost in this context implies both low first cost and low operation and maintenance cost. To accomplish these two key objectives, Thermo Electron's Cogeneration Module (TECOGEN™), has been designed utilizing components which are being produced currently in high volume. Modifications have been made where necessary to achieve the high reliability and low maintenance mandatory for this product. In addition, the system is pre-engineered and prepackaged for better quality and cost control.

The Cogeneration Module consists of a specially modified natural gas-fired internal combustion engine driving an induction generator with additional equipment for the recovery of waste heat from the engine exhaust, cooling water, and lubricating oil. The electric output is 60 kW and the thermal output is 440,000 Btu/hr. The specifications are given in Table 6-1. The energy balance for the module is shown in Figure 6-1. The electrical efficiency of TECOGEN is 26.6 percent, while overall thermal plus electrical is 83.9 percent. Figure 6-2 shows the complete cogeneration module with enclosures.

| | |
|---|---|
| INPUT: | 760 scfh of Natural Gas (1020 Btu/scf HHV) |
| OUTPUT: | Electrical - 60 kW<br>(208 V, 220-240 V, or 440-480 V; 3 phase, 60 Hz)<br>Thermal - 440,000 Btu/hr hot water<br>(18 gpm, 170°F in, 220°F out are typical) |
| EFFICIENCY: | Electrical - 26.4%<br>Combined Electrical and Thermal - 83.1% |
| DIMENSIONS: | 82 in. long x 42 in. wide x 40 in. high<br>(maximum width without acoustic enclosure - 35 in.) |
| WEIGHT: | 3000 lb |
| CONTROLS: | Completely automated via microprocessor-based control system. (Startup, monitoring, shutdown, etc.) |
| ACOUSTIC LEVEL: | 70 dBA at 20 ft |

**TABLE 6-1. TECOGEN Module Specifications.**

The cogeneration module can be represented as four interconnecting submodules as shown in Figure 6-3. Air and natural gas flowing to the Energy Generator Submodule (EGS) result in electrical power being delivered to the Electrical Interface Submodule (EIS) which, through the action of the Control Submodule (CS), connects the electricity to the external load, user's plant, or utility as the case may be.

Simultaneously, the EGS produces waste heat in the form of hot exhaust gas, engine jacket water, and lubricating oil which is delivered to the Heat Transfer Submodule (HTS) where its heat is extracted and delivered to the plant hot water load. The cold exhaust gas is rejected to the atmosphere. The CS also monitors and controls the performance of the EGS and HTS.

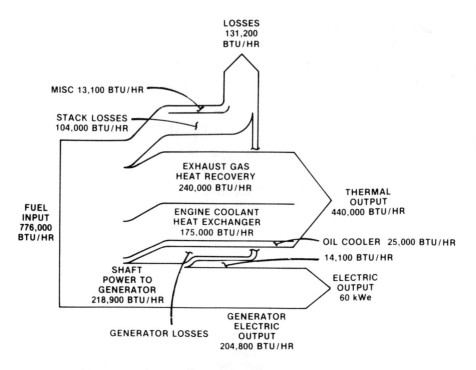

**FIGURE 6-1.** TECOGEN Cogeneration Module Energy Balance.

**FIGURE 6-2.** TECOGEN Cogeneration Module Model CM60 —
Generator End With Enclosure On.

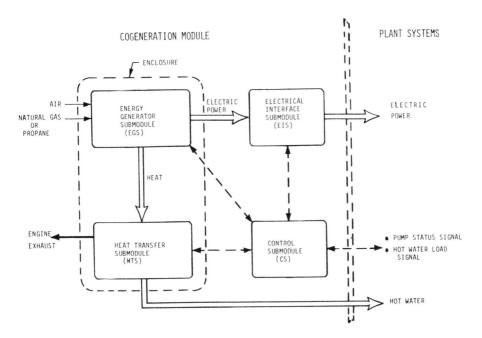

**FIGURE 6-3.** Energy and Control Presentation of TECOGEN
Cogeneration Module.

### Energy Generator Submodule (EGS)

The EGS consists of a natural gas-fired V-8 automotive engine driving an induction generator through a flywheel mounted coupling. (See Figure 6-4.)

### Engine

The engine is an 8-cylinder, 454-cubic-inch displacement automotive type. This is one of Thermo Electron's Crusader Engine Division's time-tested engines. Special modifications have been made for the TECOGEN application. These include:

- Specially developed cylinder head designed for long service intervals for exhaust valves and valve seats. An extensive research and development effort was undertaken to optimize valve and seat materials, valve angle, valve rotators, spring stiffness, and valve cooling.

- Optimized lubrication oil, oil pump, oil cooling system, and oil sump.

- Modified inlet manifold and natural gas carburetor.

- High compression ratio to take advantage of high octane number of natural gas.

- Modified cooling system to insure uniform cooling in the cylinder head.

In addition to these special design features, the engine is operated at a very modest load compared to its maximum capability so as to insure a long service life.

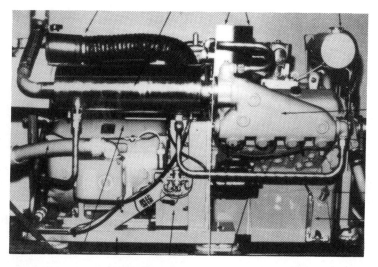

**FIGURE 6-4. Energy Generator Submodule (EGS) Shown From Right Side With Enclosure Off.**

### Generator

The generator is a simple induction unit with integral fan cooling and open, drip-proof construction. These units are much like the very common squirrel-cage motor and require no excitation system when connected to the utility lines; hence, there are no brushes or collector rings. The engine drives the generator slightly above the synchronous speed set by the number of poles and the utility line frequency (1800 rpm in this case), at which point the generator begins to absorb engine power and deliver electricity to the EIS.

### Heat Transfer Submodule (HTS)

Heat is recovered from three sources in the TECOGEN. These are the engine exhaust, the water jacket, and the lubricating oil. The colder return water from the thermal load flows first to the lube oil cooler because this is the lowest temperature heat source. It then flows to the engine jacket cooler, then to the two engine exhaust gas coolers (one for each bank), and finally through the exhaust manifolds before going to the external load. Safety devices prevent the TECOGEN from being operated unless water is flowing in this circuit. The HTS is shown in Figure 6-5.

FIGURE 6-5.  Heat Transfer Submodule (HTS).

### Exhaust Gas Heat Exchangers

The exhaust gas heat exchangers are of the water tube type consisting of a finned coil of copper tubing enclosed by an insulated stainless steel cylinder. There is also a steel baffle down the center of the finned coil. The exhaust gas is constrained to flow over the finned tubing in the annulus formed by the inner baffle and outer shell. These units are designed to drop the exhaust temperature down to the lowest level compatible with avoiding condensation and consequent corrosion problems while maintaining minimum back pressure on the engine.

The engine exhaust manifolds are also water jacketed to provide additional heat transfer from the exhaust gas. The water is directed to each of these from each of the exhaust gas heat exchangers. The heated water leaves the system after passing out of the exhaust manifolds. The water cooled exhaust manifolds also serve to reduce the amount of heat rejected to the enclosure, and ultimately to the space where the Cogeneration Module is located.

### Jacket Water Heat Exchanger

The heat given up by the engine jacket cooling system is transferred to the hot water system through a conventional shell-and-tube heat exchanger. The engine coolant (a mixture of water and antifreeze) is circulated by the engine driven pump. Safety systems are provided which prevent operation of the system in the event that coolant flow is interrupted or the coolant temperature becomes excessive. There is no thermostat in the engine coolant system.

### Lubricating Oil System

The Cogeneration Module has a specially developed engine lubrication system which serves several functions in addition to the recovery of heat rejected to the engine oil.

The engine is equipped with a very large sump (36 quarts capacity) to allow for extended periods between oil addition and/or oil changes. Normally, oil will never have to be added, but will be changed at (roughly) monthly intervals. A special oil pump with higher capacity than the standard pump and no internal by-pass is used which circulates oil from the sump to a shell-and-tube heat exchanger which transfers the heat rejected by the engine to the thermal load.

After giving up the rejected heat to the thermal load the oil is filtered and delivered to the engine for lubrication. Oil not required by the engine lube system flows back to the sump through an external relief valve. Oil pressure, temperature, and level are all monitored by the Control Submodule, which shuts the Cogeneration Module down if preset limits are exceeded.

### Electrical Interface Submodule (EIS)

The function of the EIS is to control the flow of electric power between the TECOGEN and the electrical system of the facility where it is installed. This facility will, in most cases, be supplied with power by the local utility. The EIS ensures that power produced by the TECOGEN is compatible with utility supplied power, a task made relatively simple by the use of an induction generator, and has been designed to meet the interface requirements of most U.S. utilities. The inside of the EIS is shown in Figure 6-6; the outside can be seen in Figure 6-2.

**FIGURE 6-6. Electrical Interface Submodule (EIS).**

In addition to its utility interface functions, this submodule provides a number of other safety and control-related functions such as the engine cranking control, engine ignition control, battery charging, and natural gas solenoid valve control. A redundant overspeed control which is independent of the CS is also part of this submodule. The EIS also contains a clock on the door which logs the total number of hours the system has operated and a counter which logs the number of times the system is started. (See Figure 6-2.)

## Control Submodule (CS)

The CS is a microprocessor-based system which acts as the brain of the TECOGEN. It starts the system when there is a demand for heat and shuts it down when that demand is satisfied. It senses when the EGS has reached a speed compatible with the utility electrical frequency and commands the EIS to connect the TECOGEN to the electricity network. The CS also monitors and controls the electrical power output to 60 kW through a servomotor which operates the engine throttle. The CS is shown in Figures 6-7 and 6-8.

**FIGURE 6-7. Control Submodule (CS) Shown With Door Closed.**

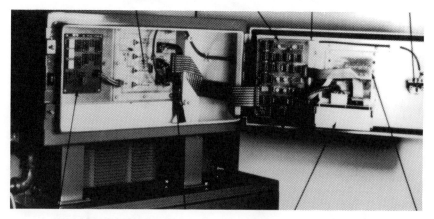

**FIGURE 6-8. Control Submodule (CS) Shown With Door Open.**

A number of operating variables can be monitored in the display area of the CS. In addition to starting, stopping, and connecting the TECOGEN to the utility, the CS monitors the output of a large number of sensors and shuts the system down if preset limits are exceeded. It also provides diagnostic capability for service personnel. The faults which are detected by the CS are listed in Table 6-2.

### ELECTRICITY RELATED

Phase Angle Error
Overvoltage
Undervoltage
Overfrequency
Underfrequency
Contactor Fault
Low Power Output
Excessive Generator Connect Time

### THERMAL LOAD RELATED

High Water Pressure
Low Water Pressure
Low Water Flow
High Water Temperature
Excessive Time to Minimum Coolant Temperature

### ENGINE RELATED

Starting Failure
Low Oil Pressure/Low Oil Level
Low Coolant Flow
High Coolant Temperature
High Oil Temperature
High Air Temperature
Overspeed
Underspeed
High Vibration

**TABLE 6-2. Microprocessor Control System Fault Detection and Display.**

### Enclosure

The Cogeneration Module is equipped with an enclosure which serves to reduce the noise level far below that of most comparable industrial or commercial equipment. At the same time, it discourages tampering by unauthorized personnel.

### Maintenance Considerations

The TECOGEN cogeneration module has been designed to offer ease of maintenance as a key feature. For these systems to be attractive to customers, it is imperative that their downtime be kept to a minimum. This is accomplished by the use of the microprocessor-based control submodule and the built-in accessibility for ease of service. The control submodule acts upon faults which may be related to electric power, thermal load, or engine by shutting the system down. It the faults are external to the system, the system will restart itself once the fault has corrected itself. In the case of faults internal to the system, the control submodule locks out the system. In all events, it displays the fault which caused the system shutdown. This facilitates trouble-shooting and restoring the system operation very dramatically.

Easy access to every component in the system minimizes the downtime during the routine and unscheduled service. Use of high-volume-based components also generally means that in most instances it is cheaper to replace a component. Because of the relatively low cost of the engine, it is planned to replace the engine at the appropriate interval (2 to 3 years, depending upon annual operating hours) rather than service it in place as is essential for heavy and expensive industrial engines. This again results in much lower downtime and improved system operation.

## COGENERATION MODULE ECONOMICS

The Cogeneration Module uses natural gas as the input fuel and produces part of its output in the form of valuable electric power and the balance as hot water. The savings generated by the Cogeneration Module primarily depend upon the differential between electric and gas rates. Since these rates vary across the nation so will the attractiveness of cogeneration. Obviously, regions with current or anticipated high electric rates and moderate gas rates are ideal for cogeneration. An example of Cogeneration Module economics is shown in Figure 6-9.

In many regions of the nation, cogenerators can take advantage of special prime mover rates offered by gas utilities. This, in effect, results in a price discount for the gas consumed by the cogenerator.

The sensitivity of the payback period to electric and gas prices is shown in Figure 6-10. It is evident that payback improves much more rapidly with the rising electric rates than falling gas rates. The effect of cogeneration gas discount also has a very significant effect on the payback. A 20-percent cogeneration gas discount improves the payback period by 15 to 30 percent, depending upon the electric rate.

The other major factor influencing the system payback is the annual operating hours. Naturally, longer operating hours would correspondingly lower the payback period. Overall a 1½- to 3½-year payback period is very feasible for many applications.

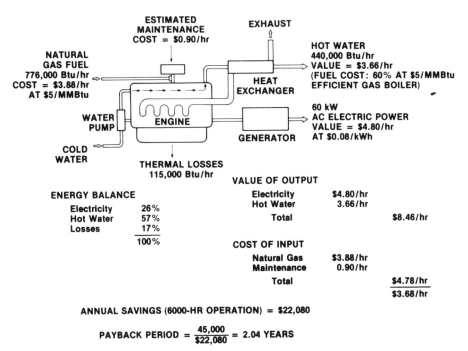

FIGURE 6-9. The Cogeneration Module Economic Advantage.

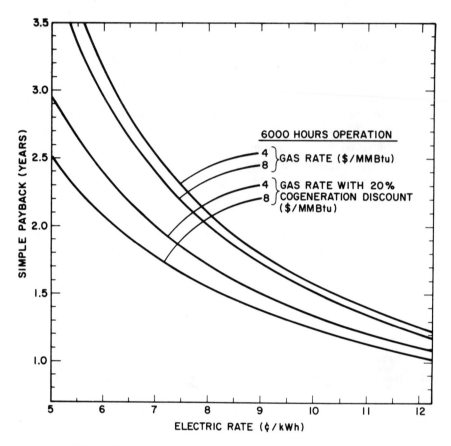

FIGURE 6-10. Cogeneration Module Payback Sensitivity To
Gas and Electric Rates.

## SYSTEM APPLICATIONS

For good economics, both hot water and electric power produced
for the Cogeneration Module have to be utilized. Whereas excess
electric power can be sold to local utilities, the hot water has to be
used on site. Good candidates for the system have to be first and
foremost large and relatively constant users of hot water or thermal
energy in other forms. This will result in long annual operating hours
resulting in short payback. Some of these operations include hospitals,
large heated swimming pools, hotels and motels, apartments and
condominiums, laundries, food and beverage processors, nursing

homes, schools and colleges, greenhouses, electroplaters, tanneries, and other light industries.

Modular cogeneration systems also offer some other unique advantages. Multiple module installations permit part load operation in response to load variation by controlling the number of modules in operation while maintaining peak efficiency at all times. In addition, these modules present a great deal of siting flexibility because of their compact size. By installing the modules in closer proximity to the load, one can minimize the losses associated with transmission of thermal energy which can be very substantial in some applications.

## CONCLUSIONS

For the first time, the benefit of economical cogeneration previously enjoyed by only large power consumers is now available to institutional, commercial, and light industrial users. With financial institutions rapidly increasing activity in creative financing, these users can share the benefit of cogeneration with little or no investment.

# CHAPTER 7
# Small Scale Systems

*S.A. Parker, W.C. Turner, R.L. Froneberger*
*C.R. Thompson*

Though research has given us many new energy technologies, the principles of cogeneration are basically the same as they were in the early 1900's, when 58% of all on-site power was cogenerated. The instances of on-site power generation, however, have declined to only 5% in 1974, and with it the amount of cogeneration.

Several factors can be cited for the drastic reduction in the number of on-site power generation facilities. One is increased government regulation of all types of electrical generation. Another is the reluctance of the utilities to provide back-up power or buy excess electricity. The biggest roadblock to cogeneration in the past, however, has hardly been something anyone would complain about—low electricity rates. Between 1940 and 1950 the real cost of electricity to an industrial user halved. As a result, it has been less expensive to purchase electricity than the equipment to produce it.

This has not been the case in Europe, where energy costs have always been high compared to the United States. As a result, cogeneration has been much more widely used, and returns have been substantial. In West Germany, for instance, where 16% of *all* power generated in 1972 was cogenerated, the average input of energy per unit of output was 38% of that of U.S. industry. The potential of cogeneration is obvious, and as energy prices continue to rise, the widespread use of cogeneration becomes imminent.

## Legal and Social Aspects

A major obstacle to widespread use of cogeneration has been legal uncertainty as to the status, rights and responsibilities of power generators. As government regulation increased, so did the fear of being classified as a utility and picking up the regulations that come from such a classification. Furthermore, there has been uncertainty and a lack of consistence over whether utilities would buy any excess electricity generated or provide back-up electricity at a reasonable price.

Both of these issues were addressed by the Public Utility Regulatory Policy Act (PURPA) of 1978. This act basically said that companies employing cogeneration to produce power primarily for their own use could not be classified as a utility, and that utilities must buy back excess electricity at rates favorable to the cogenerator, as well as provide electric backup at a reasonable rate.

What this means to small scale cogenerators is that they may proceed without fear of falling under state and federal utility regulations and may count on back-up power at a reasonable cost. This chapter will focus on small-scale situations in which the company uses all of the electricity generated.

There are environmental issues quite apart from the financial issues. Since cogeneration facilities typically extract more than twice as much useable energy per unit of fuel as standard generation units, there could be a substantial reduction in total energy requirements on a national basis with the widespread use of cogeneration. Furthermore, using less energy has the fringe benefit of decreasing the total amount of pollution from energy production.

From the non-economic viewpoint, the only major drawback to cogeneration is that since more energy is produced on site, more fuel may be burned on site and local pollution may increase. This forces the user and those who live and work nearby to bear more of the direct costs of pollution rather than those who happen to live and work near the remote power generation station. While some economists might say this is more equitable, it is likely to be perceived as a major negative factor in the decision to employ cogeneration.

## WHAT ABOUT SMALL SCALE COGENERATION?

In recent years, most interest in cogeneration has been on the very large scale, with the utility or large plant installing huge systems. With rising electricity rates, however, it is becoming more and more attractive for the user to cogenerate on a smaller scale. It is this type of installation on which we will now focus. For purposes of discussion, we will define "small scale" cogeneration as units 500 kW or less.

## Limitations

There are three main limitations to small scale cogeneration—the obvious loss in economies of scale, the need for suitable conditions (including location and rate structure), and possible strained relations with utilities.

*Economies of Scale:* It is rather obvious that to economically justify cogeneration equipment, there must be the capacity to generate and use a certain amount of heat and power. As would be expected, the cost of installing a cogeneration facility is not linearly proportional to the size of the system (that is, larger units are less expensive per kW than small ones). Of course, as energy prices continue to rise faster than the price of goods, the breakeven point relative to size will tend to come down.

*Suitability of Conditions:* Since we are talking about sites in which utility buy-back of the energy produced is not considered, there must be a match of the potential for cogeneration in terms of the power and heat produced. If all of the thermal energy produced in a topping cycle is not fully utilized, then a portion of the savings is not realized. Additionally, in the case of a bottoming cycle, competition for the waste heat as an energy source of some other form of waste heat recovery may effect the economic feasibility of cogeneration.

Physical location of the energy users and generator can influence the feasibility of a cogeneration system, as can local utility rates. Excessive heat loss through piping and storage of the thermal heat can detract from the economics. Also, as will be pointed out later in the chapter, a high electrical to fuel cost ratio will generally make cogeneration systems more attractive.

*Possible Lack of Utility Cooperation:* Although the advent of PURPA has eliminated the problem of getting back-up electricity, it may be expected that some utilities may not always be receptive to the idea of cogeneration. This issue must be considered and planned for any time cogeneration is thought of as a viable option.

## Ways Cogeneration May Be Used On A Small Scale

The key to a small scale cogeneration system is to fully utilize both heat and power. The power can be in the form of mechanical shaft power or, with the aid of a generator, electricity. When shaft power is being employed, it is typically coupled with only one machine. Electricity, however, can be wired directly to one machine, or through the utility grid. If the cogeneration system is coupled to one machine, then the savings will vary with the load on the machine. The advantage of wiring through the utility grid is that the cogeneration system can run at full load, maximizing savings, while the utility handles the remaining plant load. This is called "base loading."

In wiring the generator into the utility grid, there are three philosophies of operation, (1) Base Reduction, (2) Peak Shaving, and (3) Emergency Stand-by. Base reduction was briefly explained in the previous paragraph. When the cogeneration system runs continuously in order to minimize the amount of electrical energy purchased from the utility, it is defined as base reduction. A demand profile with base reduction is shown in Figure 7-1.

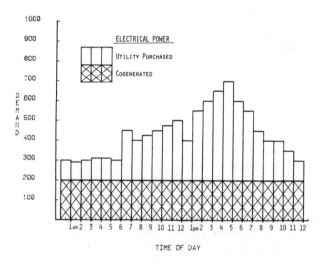

**FIGURE 7-1. Demand Profile with Base Reduction**

The next method is peak shaving. This is utilized in order to minimize the facilities peak demand. For Oklahoma and most of the South, this is especially true in the summer when a high A/C load can cause a major rachet penalty over the rest of the year. Peak shaving is done by monitoring the plant's current demand. When a set point is reached, the cogeneration system starts up and lowers the demand from the utility. Figure 7-2 shows a demand profile with the effect of demand shed.

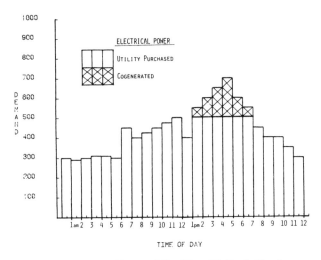

FIGURE 7-2. **Demand Profile with Peak Shaving**

The final method is emergency stand-by. In this case, the cogeneration system starts up in the event of a utility power outage. This enables the facility to continue operating. This method of cogeneration saves the facility not only avoided utility consumption but also production downtime. The demand profile with emergency power generation is shown in Figure 7-3.

In addition to the use of the power generated by the cogeneration system, there is also the thermal energy that needs to be fully utilized. The advantage of a cogeneration system over standard power generation systems is that the thermal energy (often up to 50% of the input power) is able to be utilized. The thermal energy is typically in the forms of low pressure steam, hot water or hot air. In these forms, the cogenerated energy can be used throughout a facility wherever conventional thermal systems currently exist.

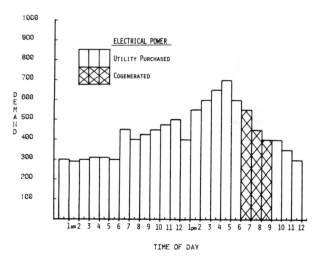

**FIGURE 7-3.** Demand Profile with Emergency
Power Generation

## OPTIONS IN SMALL SCALE COGENERATION

As the reader has probably ascertained by now, there are several different types of systems which fall under the category of small scale cogeneration systems.

### Internal Combustion Engine

The combustion engine qualifies as a cogenerator by utilizing the thermal energy in the engine exhaust and jacket cooling as well as the shaft power or electrical power of a generator is used. Combustion engines are available in a variety of sizes. Commercially available units range in size from 50 to 1350 kw. These engines can be adapted to burn a variety of fuels.

Aside from the standard #2 fuel oil, the engines can also burn #6 fuel oil, natural gas or propane. Also with the proper adjustments and set up, the engines can run on field gas, sewage gas or even coke gas. The engines are also available with a host of options which make them applicable in a wide range of environments.

As noted earlier, the power generated can be in the form of shaft power or electricity and can be used accordingly. The thermal energy, however, comes from four major sources: jacket cooling water, exhaust, radiant heat and, in some large systems, lubrication oil. Table 7-1 shows the amount of energy typically available from the various components of an engine.

| Category | Fraction of Full Load Energy Input |
|---|---|
| Shaft Power | 27 to 37 |
| Jacket Water | 28 to 35 |
| Exhaust (Total) | 30 to 32 |
| Exhaust (Recoverable) | 15 to 21 |
| Radiant Heat | 6 to 10 |
| Lube Oil Heat | 5 to 7 |

**TABLE 7-1. Typical Energy Utilization**

As with most industrial projects, the decision usually rests with a thorough economic analysis. In considering a cogeneration system using an internal combustion engine, there are four major economic considerations: (1) the resulting energy savings, (2) operating and maintenance expenses, (3) equipment and installation costs, and (4) any tax considerations. Once all of these categories have been thoroughly analyzed, an accurate economic analysis can be performed.

In order to see a range of systems, we will look at a 300 kW (case 1) and 135 kW (case 2) engine from the same manufacturer. These systems are not specifically made for cogeneration but are easily adaptable. Also, we will look at a 100 kW engine (case 3), from another manufacturer, that is specifically made for cogeneration. All three engines utilize natural gas for power and generate electricity. The difference occurs in the form of the thermal energy produced.

In cases 1 and 2, the thermal energy is in the form of hot jacket water and in the exhaust. This is where a heat exchanger can be applied to recover the exhaust heat either in the form of hot air, hot water or low pressure steam.

Case 3, however, is specially adapted to where the exhaust heat is transferred directly to the jacket water. Also, since case 3 is specially built for cogeneration, installation costs are less expensive but equipment costs are more.

Economic information for all three cases is listed in Table 7-2. Equipment and installation costs were provided by manufacturers. These cost estimates are averages, and actual costs will vary with every application. Maintenance costs were estimated by the authors. Again, actual costs will vary not only with individual applications but also from year to year. For this reason, it is important to check each application independently.

Energy information for the three cases is listed in Table 7-3. This information is easily obtainable from manufacturers. With the information listed in Tables 7-2 and 7-3, an economic analysis can be performed for all the case studies. Typically before an analysis of costs and savings is performed, you should convert the "energy recoverable" from Table 7-3 to equivalent energy saved.

This is done by dividing the energy recoverable to the efficiency of the equipment replaced (or supplemented). However, you would then take into account additional energy lost through the extra piping, if any. In these case studies, however, we will assume that the two factors cancel each other.

| Case | Size (KW) | Equipment Cost ($) | Installation Cost ($) | Investment Tax Credit | Maintenance Cost ($/Yr) |
|------|-----------|--------------------|-----------------------|-----------------------|-------------------------|
| 1 | 300 | 140,000 | 150,000 | 29,000 | 7,000 |
| 2 | 135 | 40,000 | 100,000 | 14,000 | 2,000 |
| 3 | 100 | 95,000 | 50,000 | 14,500 | 4,500 |

TABLE 7-2. Economic Information

| Case | Size (KW) | Fuel Rate (Ft³/Hr) | Operating Hours (Hrs/Yr) | Energy Recoverable Jacket Water (BTU/Hr) | Energy Recoverable Exhaust (BTU/Hr) |
|------|-----------|--------------------|--------------------------|------------------------------------------|-------------------------------------|
| 1 | 300 | 3767 | 7500 | 1,061,220 | 829,140 |
| 2 | 135 | 1750 | 7500 | 494,760 | 371,940 |
| 3 | 100 | 1250 | 7500 | 630,000 | ——— |

TABLE 7-3. Energy Information

Now that we have all information required, we will make two more assumptions: (1) the engines will run 7500 hours per year (base reduction), and (2) a 3-year payback is acceptable. With this information, a sensitivity graph of energy costs can be plotted for each case study showing the break-even line between economical feasibility and non-feasibility. Figures 7-4, 7-5, and 7-6 show the sensitivity graphs for case studies 1, 2 and 3 respectively. Again, the sensitivity graphs are a result of (1) the energy savings, (2) the operating and maintenance costs, (3) the engine's total installed cost minus the investment tax credit, and (4) a 3-year simple payback decision criteria.

Careful examination of the sensitivity graphs reveals several factors. First, and most important, small scale cogeneration is economically feasible in several areas of the country. Second, relatively high electrical costs and low natural gas costs favor cogeneration. Third, economies of scale does play a part in the economics. Fourth, the energy charge per kilowatt-hour has a more dramatic effect on economic feasibility than the demand charge per kilowatt-month.

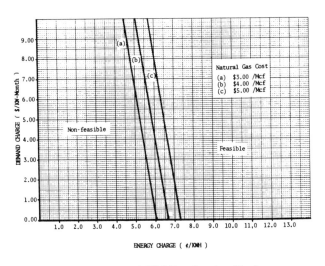

**FIGURE 7-4. 300 KW Combustion Engine**

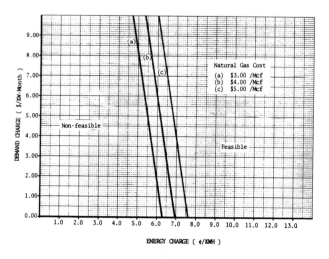

**FIGURE 7-5. 135 KW Combustion Engine**

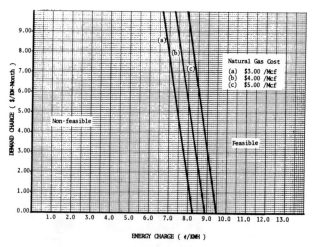

**FIGURE 7-6. 100 KW Combustion Engine**

## Organic Rankine Cycle

The organic Rankine cycle is a bottoming cycle that uses a waste heat stream as a thermal power source. Many industrial processes have low quality (low temperature) waste heat sources. Previously, there was little that could be done, economically, with a waste heat stream 150°F to 700°F. Today, however, many organic fluids exist that easily vaporize in this temperature range. Once vaporized, the organic mixture expands through a turbine generating electrical energy. The organic fluid is then condensed and recycled.

As would be expected thermodynamically, the lower the temperature of the waste heat stream, the lower the amount of power that can be generated from it, even though the hourly BTU content is equivalent. This along with the specific application can determine which working fluid the organic Rankine cycle uses.

The organic Rankine cycle can use a wide range of fluids, Ammonia, Benzene, Dowtherm A, Ethonol, Freon, Methonal, Tolene and others. Which fluid is best depends on the specific applications, however, Freon, Dowtherm and Tolene are most common. Table 7-4 gives the percentage of waste heat which can be recovered as power for various temperature ranges. There are also a wide range of sizes available for organic Rankine cycles. 100 kW to 1500 kW units are commercially available.

| Temperature Range | Type of Fluid | Percent of Waste Heat Recoverable as Power |
|---|---|---|
| 150 - 200 | Fluoro-carbon | 7 to 15 |
| 400 - 700 | Other Organics | 35 |
| 1000+ | Water | 50 |

TABLE 7-4. Recoverable Energy
Source: Polimeros (2), p. 110.

In considering an organic Rankine cycle, the same four economic considerations must be considered: (1) energy savings, (2) operating and maintenance expenses, (3) equipment and installation costs, and (4) tax considerations. Again, all of these categories have to be fully analyzed for each application in order to perform an accurate economic analysis.

Two case studies will be studied for this type of system. Case 4 is for a 300 kW system and case 5 is a 125 kW system. Case 4 utilizes a waste heat stream with a minimum of 10 million BTU's per hour according to manufacturer's data. Case 5, however, uses 500 lbs per hour saturated steam at 15 psig as a heat source. Both cases require condenser cooling water between 60°F and 70°F. The working fluids for case 4 and 5 are Tolene and Freon, respectively.

Economic information for cases 4 and 5 is listed in Table 7-5. As with the combustion engines, this information is obtainable from manufacturers. Again, the cost estimates are averages, and actual costs will vary with specific applications. Maintenance expenses, according to conversations with manufacturers, are minimal since the organic Rankine cycle are closed systems. Nonetheless, they are an important factor.

| Case | Size (kW) | Equipment Cost ($) | Installation Cost ($) | Investment Tax Credit ($) | Maintenance Cost ($/Yr) |
|------|-----------|--------------------|-----------------------|---------------------------|-------------------------|
| 4    | 300       | 370,000            | 200,000               | 57,000                    | 7400                    |
| 5    | 125       | 218,400            | 30,500                | 24,340                    | 4870                    |

TABLE 7-5. Economic Information

Table 7-6 shows the energy information for the two case studies. With the information in Tables 7-5 and 7-6 an economic analysis can be performed. Again, the holding assumptions are: (1) 7500 operating hours per year, and (2) a 3-year payback is acceptable. Figures 7-7 and 7-8 show the sensitivity graphs for case studies 4 and 5.

Careful examination of these graphs again show several enlightening points. First, and again most important, small scale cogeneration using the organic Rankine cycle is economically feasible in several areas of the country. Second, the break-even points are similar to that of the combustion engine. Third, economies of scale plays a part in the economics. Fourth, the energy charge per kilowatt-hour again has a more dramatic effect on economic feasibility than the demand charge per kilowatt-month.

| Case | Size (kW) | Waste Heat Rate ($10^6$ BTU/Hr) | Operating Hours (Hrs/Yr) | Working Fluid |
|------|-----------|----------------------------------|--------------------------|---------------|
| 4    | 300       | 10 - 15                          | 7500                     | Tolene        |
| 5    | 125       | 5.8                              | 7500                     | Freon         |

**TABLE 7-6.** Energy Information

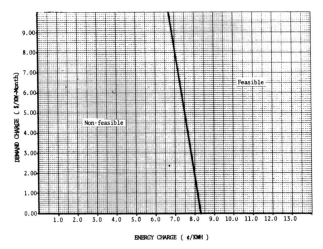

**FIGURE 7-7.** 300 KW Organic Rankine Cycle

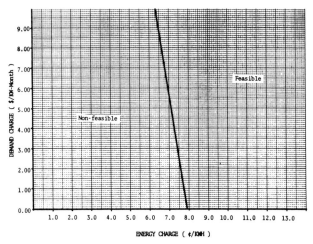

**FIGURE 7-8.** 125 KW Organic Rankine Cycle

At the time of this writing, the authors had not completed the economics analysis for small scale cogeneration systems involving steam Rankine cycles or gas turbine systems. However, the principles of the systems can be partially explained.

### Steam Rankine Cycles

In many facilities, low pressure saturated steam is used either in process or for heating. In a topping cycle, the boiler could be used to produce high pressure superheated steam instead. This could be expanded through a turbine to produce power, then the low pressure saturated tailings from the turbine could be used as before. The logic here is that the energy generated is more valuable than the incremental fuel consumed.

If a facility had a high temperature waste heat source, it could use a bottoming steam Rankine cycle. Similar to the bottoming organic Rankine cycle noted earlier, the waste heat could be used to generate superheated steam. This is turn could be expanded through a turbine to generate shaft power or electricity. The saturated steam tailings could then be used elsewhere or condensed and re-cycled. Savings are best when the steam tailings are put to additional use.

### Gas Turbine System

In the gas turbine system, energy is produced by combusting a fuel in compressed air and allowing the exhaust gases to expand through a turbine to produce power. The hot exhaust gas from the turbine, typically around 1000°F to 1400°F is used to heat water, produce steam, or heat air through a heat exchanger.

This method can be used when the exhaust gas is a more convenient heat source than steam, or when there is a sporadic need and bringing steam to pressure is a problem.

### WHAT THE FUTURE HOLDS

Although cogeneration is an old technology, it is evident that there is currently much room for growth. That growth largely depends on three factors—energy cost, and energy supply and legislation.

## Energy Costs

As the reader has probably perceived, there is a very simple relationship between energy cost and the economic feasibility of cogeneration. As the cost of electrical energy goes up relative to combustion fuels, the attractiveness of cogeneration increases dramatically. This is driven by the fact that electricity is usually purchased while heat is usually produced on site by combustion. Therefore, the savings of cogeneration is the result of the avoided cost of electrical purchases, while fuel cost must be treated as an operating expense.

Although it is dangerous to try to predict what energy prices will do in the future, recent developments do indicate a trend that may continue for the next several years. As consumers responded to high oil prices by learning to reduce energy consumption, the recently deregulated oil market experienced the predictable price drops. This fact coupled with the weakening of OPEC's price setting structure have set the stage for lower oil prices than have been seen in recent years. Natural gas prices may also become more responsive to market forces.

## Energy Supply

The second factor affecting cogeneration in the future is the supply of fuels used to generate electricity and heat. While there is general concensus that the supply is shrinking, estimates as to how long the supply of the different resources will last vary. The diminishing energy supply has two effects on cogeneration.

The first factor—rationing—has a negative effect on cogeneration. Since topping cycles and (indirectly) bottoming cycles depend on fuels as a primary energy source, the prospects of a cut-back or rationing of fuel could seriously deter cogeneration.

The second factor is probably much more important than the first one. Since cogeneration is at least twice as efficient as simple electrical generation, the employment of cogeneration could have substantial positive impact on the amount of total fuel used. From a macro-economic point of view, this makes cogeneration very attractive.

**References**

*The Industrial Cogeneration Manual,* The Regulatory Policy Institute, Washington, D.C., 1983.

Polimeros, George, *Energy Cogeneration Handbook*, Industrial Press, New York, 1981.

Turner, Wayne C., *Energy Management Handbook*, John Wiley and Sons, New York, 1982.

Various trade literature.

Wilkinson, Bruce W. and Barnes, Richard W., *Cogeneration of Electricity and Useful Heat*, CRC Press, Boca Raton, Florida, 1980.

# CHAPTER 8
# A Prepackaged
# Gas Cogeneration System

*J.R. Clements*

In September, 1982, United Enertec, Inc., and Martin Cogeneration Systems were contracted by GRI for Phase I of the Research and Development of a Packaged Cogeneration System for hospitals. This project was initiated as part of GRI's near-term effort to develop packaged cogeneration for early market entry while working on long-term component improvement.

Martin Tractor Company was selected in its capacity as a Caterpillar dealer with over 50 years in the general engine business, as well as 20 years of experience with cogeneration/total energy systems. In addition, Martin has been involved in developing a generalized cogeneration package for the last 2 years with its subcontractors, United Enertec, Inc., of Dallas, Texas, and Energy Management and Control Corporation of Topeka, Kansas.

The objective of Phase I was to design and evaluate a pre-engineered, packaged gas-fired cogeneration system for mid-sized hospital facilities and to assess the potential thereof. The principal field investigators were James R. Clements and James R. Macanliss of United Enertec, Inc. The chief design engineer was Doug Wallace, P.E., of Energy Management and Control Corporation. Overall project control was maintained by Harry W. Craig, Senior Vice President of Martin Tractor Company.

The following tasks were accomplished in the Phase I effort:

*Application Analysis:* The entire hospital market was surveyed through data provided by the American Hospital Association to determine that over 4,000 of the 6,965 U.S. hospitals are potential candidates for a single-unit package installation. Average hospital energy loads were determined on a thermal, electrical and cost basis.

Specific energy use data was obtained from over 400 hospitals, primarily those operated by Hospital Corporation of America (HCA) and the Veterans Administration (VA). On-site audits of 15 HCA facilities were conducted and daily, weekly, monthly and yearly energy profiles were developed.

Excellent cooperation was received from HCA, which presently owns and operates over 350 "for profit" hospitals world wide and has 100 additional facilities under some stage of construction. HCA has demonstrated extreme interest in the results of this study and of cogeneration packages, and HCA is the largest in the world. Other major holding companies were contacted and exhibited similar interest.

In the on-site audits, siting and operational constraints were evaluated. Although certain facilities will have more favorable characteristics in these areas, no unresolvable obstacles were identified.

Many of the facilities surveyed utilized four-pipe heating and cooling systems which are ideal for a cogeneration package. In this situation, the hot water BTU's from the system can be directly injected into the hot water line practically eliminating the need for a boiler to produce domestic hot water and space heating. The cold water output from the absorption air conditioning can be injected into the cold water line. In most cases, unused capacity in the existing cooling tower can be used for the absorption unit.

All of the facilities audited housed existing emergency generators of 300+ kW capacity. These units would be useful to avoid demand charges when the cogeneration unit is down for maintenance.

Each of the facilities had space for the package in the general proximity of the existing power room. Especially noise sensitive areas did not appear to present a problem for the designed package noise level.

Each facility appeared eager to do something to reduce its spiraling energy costs.

*Module Design and Performance Analysis:* A general design of 300-450 kW capacity was identified to provide the widest application for the hospital industry in general. Design and analytical studies were accomplished to determine optimal cost/performance characteristics for each component and the entire system. A detailed design was completed and alternative component manufacturers were identified and contacted.

An isochronous generator was selected to parallel with the electrical utility in order to provide additional emergency backup, as well as provide power factor control.

Heat recovery in the form of 210° hot water was chosen as the basic thermal medium; however, a steam separator can be easily added to provide 40 to 80 PSI steam if required.

A housing and base 38' wide and 10' high was chosen to minimize cost and area required. The control system was designed to provide all prealarm, alarm and automatic shut down capabilities. In addition, it was designed to sense electrical and thermal requirements to operate the system at full load only when the thermal output can be utilized.

*Component Selection:* Numerous component manufacturers were contacted and evaluated on cost, reliability and service capability. Excellent cooperation and interest were received from the manufacturers.

The entire system was divided into the following subsystems:

1. Prime Mover
2. Generator
3. Switchgear
4. Control System
5. Heat Recovery
6. Housing and Base
7. Absorption Air Conditioning

Component selections were based on current cost and current technology; however, further technical improvement and cost reduction opportunities were identified.

*Commercial Feasibility Assessment:* A costed bill of materials was prepared based on current OEM costs, predicted mass-production costs and predicted improved technology costs. The range of these costs varied by over 25%. Maintenance costs and other operating costs were identified. The operation costs characteristics were integrated with the system's anticipated performance data to determine the financial results of installing a system in five separate markets.

The results varied widely due to specific utility rate structures with predicted annual savings ranging between $103,000 and $271,000. An analysis through 1995 was made using predicted escalation of fuel and electricity rates. All markets indicated adequate return on investment ranging from a 1.5- to 5-year payback.

Benefits of the system to the utility rate payer, gas industry and the hospitals were identified. As the 450 kW system consumes 49,056 MCF per year of gas at full load, large commercialization could lead to the consumption of $198 \times 10^6$ BCF per year. In addition, the normally weak demand for gas during the summer months would predictably be the highest usage of the system. The increased gas sales over a year profile would have the effect of stabilizing the gas utility load and reducing the consumer cost.

A summary of the system's anticipated financial performance follows:

### Capital Cost of 450 kW package - installed*

| Present Cost | Mass Production Cost | New Technology Cost |
|---|---|---|
| $464,490 | $402,000 | $345,000 |
| $1,032/kW | $890/kW | $700/kW |

*These values include profit to the assembler/marketer/general contractor based on an estimated appropriate margin.

In order for the purchaser of a Hospital Cogeneration Package to pay back this substantial investment, the load profile and the utility rate characteristics must meet certain criteria. The hospital and energy load demand must utilize the full thermal and electrical output of the package for at least 6,000 full-load hours. In addition, the differential between the price of the fuel purchased and the electricity displaced must be equivalent to $.05 per kwh ($15 per MMBTU) generated.

For the prototype installation, the site selected will be in an area that has gas priced at no greater than $5.00/MCF which is roughly equivalent to $0.0625/kWh generated and electricity at no less than $.08/kWh. These locales are quite common in California and New

York, and they are increasingly occurring throughout the rest of the U.S.

Under a scenario of $5.00/MCF gas, $0.08/kWh offset electricity, and offset thermal requirements at the $5.00 rate and 75% efficiency of the alternate heat source, the following economic performance will be anticipated for the testing period:

Electricity Cost Saved                                      = $272,000
(8,000 HRS) (425 Net kW) ($0.08/kWh)

$$\frac{\text{Thermal Energy Saved (assuming 100\% utilization)}}{0.75}$$
(8,000 HRS) (2.972 MMBTU/HR) ($5.00/MMBTU)     = $158,506

Cost of Gas                                                 = ($224,000)
(8,000 HRS) (5.60 MMBTU/HR) ($5.00/MMBTU)

Maintenance                                                 = ($  18,000)
($.005/kWh) (8,000 HRS) (450 Gross kW)

Predicted Savings                                           = $188,506

In many instances, the hospital's thermal requirements are not continuously sufficient to utilize all of the system's thermal output. Therefore, in order to utilize the thermal energy year-round, 72 tons of absorption air conditioning have been added to the package for the equivalent of 5,000 full-load hours per year.

During the warmer months, it is anticipated that only half of the thermal output can be utilized for domestic hot water and limited space heating, thus absorption air conditioning utilizing 46% of the output has been included. This tonnage can be reduced or increased depending on the site specific requirement. The cost associated with this change would be approximately $900/installed ton.

With the absorption air conditioning added, the financial results become:

Electricity Generated Savings                              = $272,000
(8,000 HRS) (425 Net kW) ($0.08/kWh)

A/C Electricity Savings                                    = $36,000
(5,000 HRS) (72 tons) (1.25 kW/t) ($0.08/kWh)

A/C Thermal Energy Utilized                                    = 6,840 MMBTU
(5,000 HRS) (72 T) (19,000 BTU/T)

Thermal Savings                                                = $112,907

$$\frac{[(8,000 \text{ HRS})(2.972 \text{ MMBTU/HR})-6,840 \text{ MMBTU}]\,(\$5.00/\text{MMBTU}}{0.75}$$

Cost of Gas                                                     = ($224,000)
(8,000 HRS) (5.60 MMBTU/HR) ($5.00/MMBTU

Maintenance                                                    = ($ 18,000)
($.005/kWh) (8,000 HRS) (450 Gross kW)

Predicted Savings                                              = $178,907

Thus, the net savings will be decreased by $9,599; and, the capital requirement will be increased by approximately $50,000. With the installed target of $900/kW, the cost of $405,000 (450 kW) ($900/kW), would be paid back in 2.28 years.

*Conclusions:* The system's financial performance has been calculated at optimal system performance with favorable utility rates. However, if this performance can be attained or improved upon, and if the improved system cost can be accomplished, the Hospital Cogeneration Package should have outstanding market potential throughout the country.

## PHASE II

The conclusion of Phase I effort is that a cost effective cogeneration package can be developed for a significant number of hospital facilities.

The purpose of Phase II will be to confirm the assumptions that lead to the Phase I conclusion, as well as to point out specific research that may lead to improved system cost and performance. Several of the key variables to be confirmed in Phase II are:

1. First Cost of Hardware        4. Reliability
2. Installation Cost             5. System Efficiency
3. Maintenance Cost             6. Actual Savings

Each of the above must meet specific goals for the system to be commercially viable. To determine whether this is possible, a prototype system will be built and each factor tested. The prototype will then be installed in an actual hospital and operated for a least one year.

Phase II is expected to take approximately 24 months from inception to installation and monitoring. The following steps will be accomplished:

1. Monitoring and selection of a hospital before installation in order to establish pre-cogeneration operating characteristics.

2. Developing detailed production drawings and building the prototype.

3. Testing the package in a controlled environment at the factory.

4. Installing the prototype at the hospital.

5. Monitoring and evaluating specified variables and operating results.

As the data is gathered, compiled and analyzed, a commercialization plan will be developed. The Caterpillar dealer network will be utilized for marketing, installing and servicing the system.

In summary, Phase II will confirm the system's capabilities, identify any obstacles to be addressed and provide a base for commercialization.

The successful development of the Hospital Cogeneration Package would produce significant sales of equipment and natural gas. Phase I indicated the potential for as many as 5,000 hospital installations. Each unit would consume 44,800 MCF of natural gas per year, less a potential maximum reduction of 26,298 MCF by thermal recovery, leading to a net increased consumption of 18,502 MCF/year. Thus, the increased gas consumption would total 83,259,000 MCF/year. However, this increased gas consumption is understated by the amount of thermal energy that offsets electricity (such as air conditioning) or fuel oil (certain boiler applications). It is also anticipated that modifications to the Hospital Cogeneration Package would have significant application for nursing homes, food processing plants, and numerous other operations requiring hot water or low pressure steam. These markets would open up thousands of additional opportunities for increased equipment and natural gas sales.

# CHAPTER 9
# Third Party Ownership and Operation
# of Cogeneration Facilities

*W.G. Reed, R.H. McMahan, Jr.*

## ECONOMIC BENEFITS OF COGENERATION

Let's review the basic economics of cogeneration.

Suppose you have an industrial processing plant that uses steam at a rate of 400 thousand pounds an hour. This could be a 150,000 barrel-per-day refinery, or a medium-sized integrated paper mill, or perhaps a typical chemical processing plant.

Steam is produced in a gas-fired boiler consuming some 4 billion cubic feet of gas per year (Figure 9-1). With gas currently at $3.50 per 1,000 cubic feet, and assuming a 6% inflation rate, the annual cost of operating the boiler would be 16 million dollars in 1986.

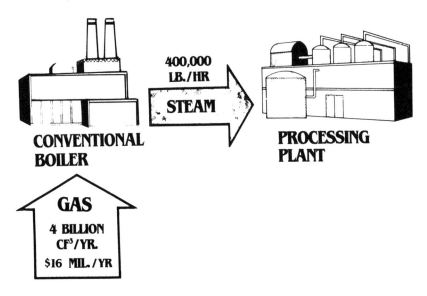

FIGURE 9-1. Typical Industrial Steam Production.

133

Suppose you replace that boiler with a gas-fired cogeneration system, at a cost of 70 million dollars. If you had started work in 1984, you could figure on start-up early in 1986 (Figure 9-2).

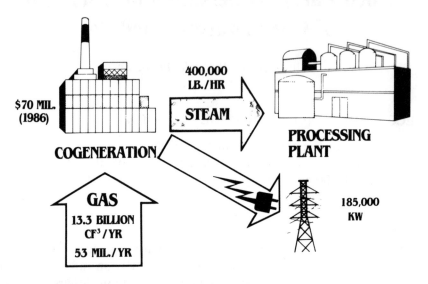

**FIGURE 9-2. Industrial Boiler Replaced by Cogeneration.**

The cogeneration system produces steam at the same rate as the old boiler, but, in addition, produces 185,000 kilowatts of power. You sell this power to the local utility at a current rate of 5 cents per kilowatt hour, producing an annual revenue of 70 million dollars in the first year of operation.

The amount of fuel consumed increases to 13.3 billion cubic feet a year, costing $53 million in the first year. Financial results for that first year of operations are shown in Figure 9-3.

- You receive 70 million dollars from the sale of power
- You save 16 million dollars by not operating the old boiler
- Your total expenses for operating the new cogeneration system are 59 million dollars, including 53 million for fuel
- Your net pre-tax cash flow from operations is 27 million dollars— roughly a 2½-year payback.

Over 10 years (Figure 9-3), assuming annual escalation at 6%, the combination of power sales and steam cost savings provides one billion, one hundred twenty-eight million dollars. Plant operating costs take 774 million, leaving a net cash flow from operations of 157 million dollars.

## FINANCIAL RESULTS
### $ Millions

|                          | 1986   | 1986-95 |
| ------------------------ | ------ | ------- |
| Power Sale Revenues      | $  70  | $ 919   |
| Cost Savings             | 16     | 209     |
| Total Benefits           | $  86  | $1128   |
| Cost of Operation        | 59     | 774     |
| Net Cash Flow (Adjusted) | $  27  | $ 157   |
| 10-Year IRR              |        | 25%     |

FIGURE 9-3. Results of Typical Cogeneration Application.

This amounts to an internal rate of return, over 10 years, of 25 per cent—an attractive investment in most industries.

Of course, these results will vary for different conditions (Figure 9-4). For example, for steam loads between 100,000 and one million pounds an hour, and for 1986 electric power sale prices from 4.5 to 5.5 cents, the Internal Rate of Return could range from under 15% to nearly 40%.

Potential returns from cogeneration can be very attractive. The key question is: if all these benefits are available in a cogeneration project, how can a company go about capturing them?

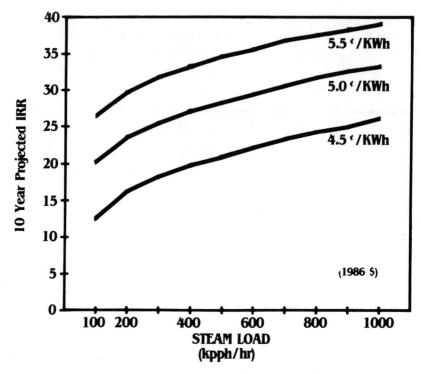

FIGURE 9-4. Variation of Project Returns with Steam Load
and Electricity Sale Price.

## BUSINESS ISSUES IN
## IMPLEMENTING COGENERATION PROJECTS

Let's suppose your company has gone through this sort of analysis
and determined that cogeneration could provide big benefits. You've
done enough engineering to know generally what kind of system you
need; the issue now is how to get it. What sort of financing and
business arrangements will best fit your circumstances?

To answer this question, let's examine the criteria your management might use to evaluate alternative business deals.

First, you must consider your financial objectives. These would
include . . .

— The requirement to utilize capital resources wisely, considering
   the cost of capital as well as alternative uses for it.

- The need to maintain adequate capitalization ratios, thereby avoiding adverse credit ratings and maintaining the integrity of your balance sheet.

- A desire to obtain high levels of profit, both in *absolute* terms, and, perhaps even more importantly, . . .

- In terms of the *rate of return* on your capital.

In striving to meet these financial goals, you want to select projects that present minimum risks. This will lead you to look for other characteristics.

- You'll try to avoid dilution of critical resources. You'll want to keep your people available for your mainstream business.

- a related point would be a desire to "stick to the knitting"—to avoid getting entangled in an unfamiliar activity, such as the operation of a power plant, where you might not be as effective as you'd like. This desire, however, would have to be balanced against . . .

- a desire to retain control of your steam supply, which is critical to your process operations.

- Lastly, you may want to minimize the risks associated with power sales to utilities (the risk that electricity buyback rates may fall), and fuel supply contracts (the risk that the cost of fuel may increase or its availability decline).

Your assessment of all these considerations will influence the way you might structure the ownership and financing of a cogeneration project.


## OWNERSHIP AND FINANCING OPTIONS

Options for owning and financing a project might be broadly categorized in three general ways:

- You could own and finance the project yourself, using your own equity or debt funds.

- Alternatively, another party could provide lease funds while you operate the plant.

— Finally, a third party could do it all—could finance, build, own, and operate the plant for you. This third party could even be a joint venture or partnership in which you participate, in order to secure more of the project benefits for yourself.

(Incidentally, let me anticipate a question that sometimes arises about this term, "third party." If we're just talking about you, the industrial, and someone else who owns the cogeneration plant, why is he a *third* party? It has to do with the effect of PURPA and the fact that most cogeneration projects will be selling power to the local utility. In this situation, you, the industrial, are the first party. The utility is the second party. So if someone else owns and operates the cogeneration plant, he becomes the third party in that arrangement.)

Your choice among these ownership and financing options will depend on how they meet the criteria we just discussed. Let's see where each of these approaches generally satisfies these objectives, where it works against them, or where the impact may vary, depending on your particular circumstances (Figure 9-5).

The first objective was to minimize capital cost, or preserve capital for alternative uses. A self-owned project would require that you use your own resources—your cash or credit—so we show a negative for that objective. But the lease and third-party options get a plus, because none of your capital is required.

As for balance sheet impact, if you use your own funds, the asset would appear on your balance sheet, and this might adversely affect your credit rating. A lease would not be on your balance sheet, but might require a footnote. A third-party arrangement may have no impact at all.

The *absolute value* of dollar returns is greatest if you finance the project, because you don't have to share the benefits. If someone else is involved—a lessor or a third-party owner-operator—he will take a portion of the project returns, leaving less for you.

The *rate* of return, however, can be considerably high in the lease or third-party cases where you don't have your own funds invested.

If you design, build, own and operate the plant yourself, a lot of your people and funds will be tied up for a long time. A lessor takes some of this burden by providing the financial resources, but you must provide for operation. The third-party option relieves you of all these burdens.

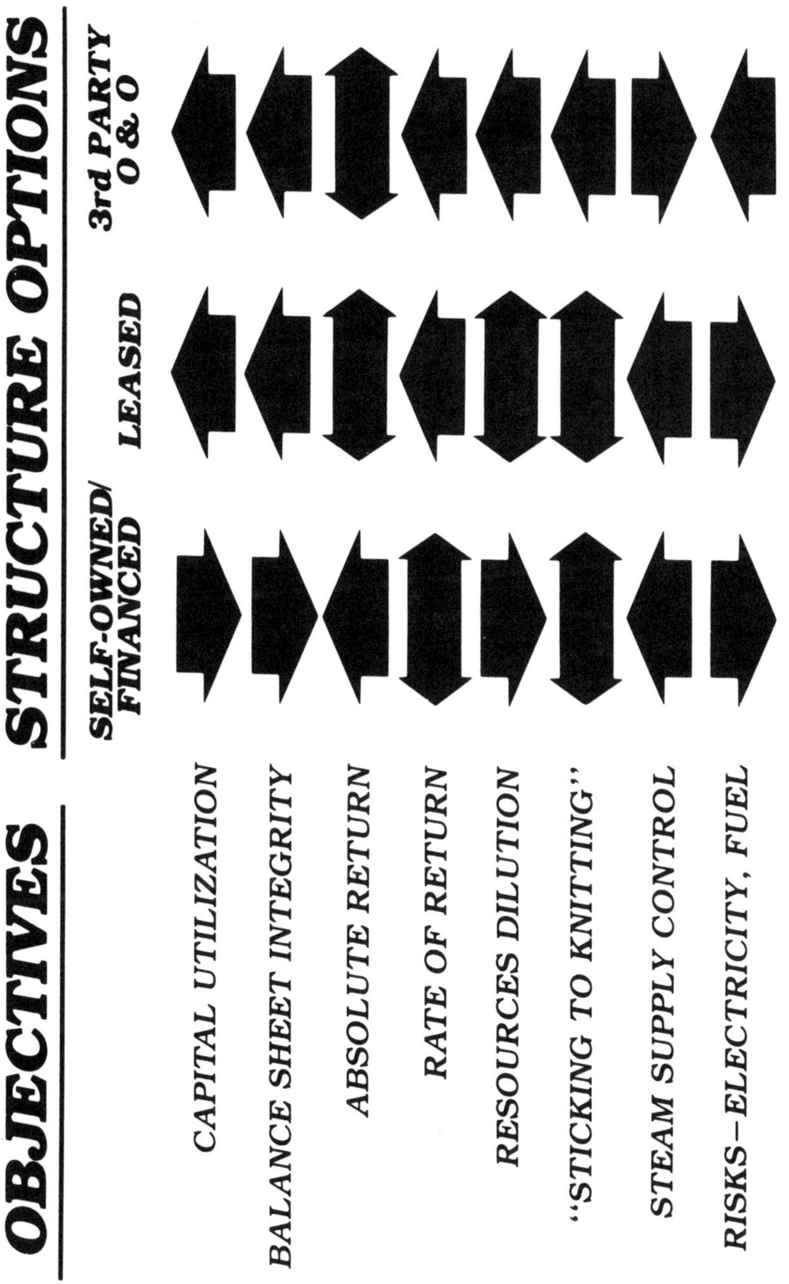

FIGURE 9-5. How Alternative Ownership and Financing Arrangements Meet Financial and Business Criteria.

Power plant operation may or may not be familiar to you. So, in those cases where you operate the plant, this rating could go either way. But where the third-party owner operates the plant, you avoid this chore completely.

If you own or lease, you operate the plant yourself, so you have direct control of your steam supply. In the third-party case, you are dependent on the operator for your steam. The way to minimize the risk in this situation is to deal with an experienced, reliable third-party operator.

As for fuel and electricity risks, we show negatives for self-ownership or leasing, because you must negotiate fuel supply and the electricity sale rates, and you bear the risk of future changes. But a third-party owner-operator can assume all these risks. You may wish to retain some of this risk yourself, in order to increase your returns. It is the allocation of these last two risks, along with the responsibility for successful operation of the plant, that principally govern the sharing of project benefits.

This is an over-all comparison of how these three financing structures satisfy your business and financial objectives. Each offers some rewards; each has some attendant risk. You might judge the value—the importance—of each criterion relative to the others. No one approach is right for everyone, but many companies are choosing to have a third party own and operate their cogeneration facility.

## THIRD-PARTY OWNERSHIP

To see why, let's go back to the economic example and carry it one step further.

In that example, you could realize a 25% return from a cogeneration investment of $70 million. But, suppose that you also need to upgrade your plant's production capabilities to remain competitive. In fact, your plans call for a $70 million expansion, and the economics of your business are such that this investment will yield a 20% return. But capital resources are limited—you can't finance both the cogeneration and the plant expansion.

Clearly, you should make the 25% cogeneration investment, and forego your 20% process improvements, right? (Figure 9-6.) "No way!", say your production and marketing people, "We've got to

# INVESTMENT OPTIONS
## $70 MILLION

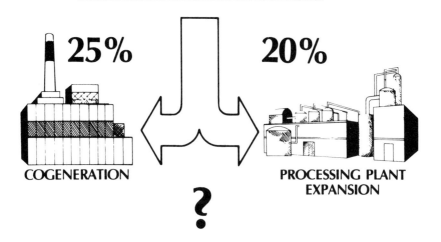

**25%**          **20%**

**COGENERATION**          **PROCESSING PLANT
EXPANSION**

**?**

FIGURE 9-6. Competition for Capital Can Complicate
Cogeneration Investment Decision

expand!" A difficult choice. Without both, you may lose ground to
your competitors.

The fact is, you may be able to *have* both, without exceeding your
capital budget.

Suppose you can find someone else—that third party we talked
about—who's willing to build and operate the 70-million-dollar
cogeneration facility for you (Figure 9-7). He will, of course, want
to share in the returns, but because he is familiar with power generation technology, he will be satisfied with a 15% return on his investment. This leaves a 10% return for you, on an investment you didn't
make. You put your 70 million dollars into your process. You
improve your competitive position, and realize a combined return
of 30%: 10% from the cogeneration, and 20% from the process
improvements. Not a bad solution.

And remember, you can realize savings like this while at the same
time preserving capital for other uses, keeping resources free for
other activities, and avoiding the complexities of negotiating fuel

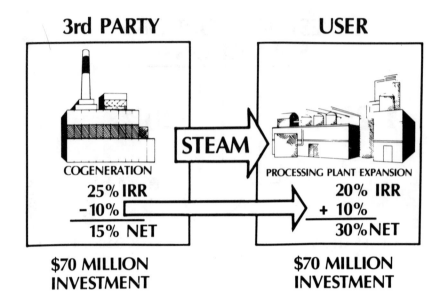

**FIGURE 9-7.** Third-Party Ownership of Cogeneration Facility
Can Increase Returns on Your Financial Investments.

supply and power sale contracts. All you have to do is share in the project benefits.

How big a share of those benefits you receive will depend on how much of the project risk you chose to take.

## A.  Approaches to Risk and Benefit Sharing

There are many ways of allocating the risks and benefits between the industrial host and the third-party owner-operator. One of the more common arrangements, the "steam-discount approach," places full responsibility for fuel supply and electricity sale, as well as for plant operation and maintenance, on the third party (Figure 9-8). In this case, he is, in effect, your "over-the-fence" steam supplier. You receive your share of project benefits in the form of a discount in the price you pay for steam—a discount from what you would have paid to produce that steam in your own boilers.

But suppose you want to increase your share of benefits by taking on more of the project risks. One way to do this is to take on the fuel contract and handle the sale of power to the utility (Figure 9-8).

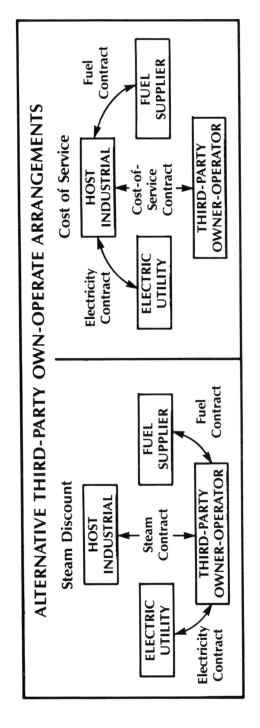

FIGURE 9-8.  Choice of Own-Operate Arrangements is Determined by Benefit and Risk Shares Desired by Host Industrial.

Under this arrangement, you become responsible for all of the inputs to, and outputs from, the cogeneration plant. The third party provides a service: he operates the plant to convert your fuel into your steam and electricity. For that service, he receives a fee—what we call a "cost-of-service" fee—based upon the output of the plant. By assuming the fuel supply and power sale risks, you substantially increase your share of the rewards.

There can be many variations on these arrangements. So far, we've yet to see any two projects that look exactly alike.

## B. Implementation

Whatever project structure is worked out for allocating benefits and risks, it must be implemented by specific contracts and agreements among all the parties involved. The exact form of these agreements will vary, but the issues they must deal with can be pretty well identified. Whether these issues are covered in one overall agreement, or a series of separate contracts, here are some of the things you'll have to resolve:

First, there would be an underlying agreement between the host industrial and the owner-operator that defines how the risks and rewards will be shared. In some cases, the host industrial may elect to participate directly in the project by becoming a joint-venture partner with the third-party operator. In any event, this basic agreement will specify the duration of the relationship, and describe what happens at the end of that time.

There will be a contract that specifies the amount of steam the cogenerator agrees to deliver and that you agree to take. It will define your right to curtail steam deliveries. It will also designate responsibility for backup supply when the cogeneration plant is not operating.

This contract will specify the price you will pay for the steam, if it's a steam-discount arrangement; or how the operator's fee will be determined, if it's a cost-of-service deal.

There will probably be an agreement whereby you lease the site for the facility to the owner-operator.

There may also be a contract under which you provide services to the cogeneration facility, such as water supply, waste disposal, or even administrative support.

Each project is unique, but there are lots of details to be arranged.

## C.  Examples of Third-Party Projects

So far, we've discussed concepts. Now, let's look at some real examples of own-and-operate projects that are underway or proposed, to see how these concepts work out in practice.

The first example is a project currently under construction, and due to start operation in 1984 (Figure 9-9).

Our steam host had capital to invest and wanted substantial participation in project benefits. So, a joint venture was formed which will own and operate the cogeneration plant; GE and the host provided equity funding, and will share net returns.

GE will sell a turnkey power plant to the joint venture, and provide operational and maintenance services. The steam host will lease the plant site to the JV, and provide support services under contract.

The host is responsible for providing fuel. In this case, GE, through its subsidiaries, Ladd Petroleum and American Pipeline Co., was able to provide an attractive price for long-term fuel supply that will provide a major share—although not all—of the project's needs. The host has contracted with other sources for the balance of the fuel.

The host purchases the fuel, which he then provides at no charge to the joint venture. In return, the host receives free steam.

The JV sells its electrical output to the local utility, and passes the revenues through to the host. Thus, the host is bearing the risk of fluctuations in the price the utility pays.

The host pays the joint venture a cost-of-service fee which is proportional to the amount of power sold.

So, in summary (Figure 9-10), this project involves joint venture ownership, equity funding, and the cost-of-service business structure. The special feature is that GE supplies much of the fuel.

In the second example (Figure 9-10), GE would be the 100% owner-operator, with financing arranged through a lease from the General Electric Credit Corporation. The customer is a refinery looking for a market for low-Btu gas, which is a by-product of its operations. We have proposed to mix this by-product gas with natural gas and use the blend to fuel a cogeneration plant. We would pay a price for the low-Btu gas that is more than the refinery could get on the open market, but less than the price of natural gas.

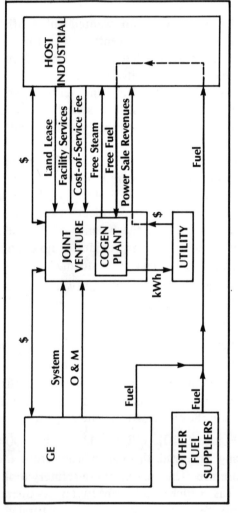

FIGURE 9-9. Structure of an Existing Cost-of-Service Project.

## KEY CHARACTERISTICS OF THREE OWN-AND-OPERATE PROJECTS

| | Case 1 | Case 2 | Case 3 |
|---|---|---|---|
| OWNERSHIP<br>FINANCING<br>SPECIAL FEATURES | • Joint Venture: GE & Host<br>• Equity<br>• Cost-of-Service<br>• GE fuel supply | • GE<br>• Lease<br>• Host gets premium price for by-product fuel | • Joint Venture: GE & AE<br>• Lease<br>• Joint Venture operates existing facility<br>• Host shares Joint Venture revenues |

FIGURE 9-10. Examples Illustrate the Diversity of Arrangements that is Possible in Own-and-Operate Cogeneration Projects.

The customer receives his benefits through the sale of his by-product fuel at a premium, while the owner-operator's benefits come from the use of low-cost fuel to generate the electric power it sells to the utility.

In another case (Figure 9-10), an architect-engineering firm had performed the initial cogeneration economic studies which showed their client the value of cogeneration. GE and the AE have formed a joint venture to build, operate and maintain a cogeneration plant, financed through a GECC lease. The joint venture will also lease the customer's existing energy facility, and operate it jointly with the new cogeneration plant.

The host in this case is a tax-exempt institution. He was concerned that if he shared directly in project benefits, such as through a steam discount, his returns might be considered taxable as "non-related business" income. Therefore, this project is structured so that the host buys steam and power from the joint venture at full market prices. He received his returns through the land lease, where payments include a percentage of joint venture gross revenues.

The key point in these examples is that third-party ownership can work in a wide variety of situations, but each project must be carefully analyzed and structured to serve the specific needs of the individual customer.

The only way to reach a satisfactory result is to sit down with a responsible owner-operator and describe fully your business goals and constraints, as well as your technical requirements. If you've picked the right party to talk to, he should be able to help you define not only the best system to meet your technical needs, but also the best ownership and financing arrangement to serve your business objectives.

## SELECTION OF A THIRD-PARTY OWNER-OPERATOR

There are a lot of people offering to play the role of cogeneration owner-operator these days: architect-engineers, fuel supply companies, independent entrepreneurs, utilities, and equipment suppliers like GE. On more than one occasion, a project has been announced with high hopes for large benefits to the industrial, only to fall apart in a few months because some part of the proposed deal couldn't be worked out.

Here is what you should look for in an owner-operator who can put together deals that work.

A responsible owner-operator must have the technical capability to analyze the economics of your project, and select the best system for your needs. And he must have the experience to assure that those analyses are realistic.

He should understand utility economics and system operations, since these can have a great impact on project returns.

He should be experienced and flexible in structuring business arrangements.

He should have established access to a variety of financing sources.

He should be able to arrange a fuel supply on advantageous terms.

He should be able to coordinate design, equipment, construction, and installation to deliver a reliable system.

And he should have the personnel and facilities necessary to operate and maintain the plant for maximum availability over its entire life.

Finally, he should recognize that, for long-term success, the project he designs must serve the needs of all concerned parties: the cogenerator, the utility, the ratepayers, and you, the industrial. These interests sometimes appear to conflict. But the benefits can be large, and if the project is structured properly, everybody can win.

# CHAPTER 10

# Cogeneration Planning: Industrial and Third Party Perspectives

*G. Decker*

The last economic recession had its impact on cogeneration in the form of deferral of many projects. Demand for products was down, capital was scarce, and interest rates were going up. Many potential industrial cogenerators became very cautious, and this caution has persisted into the present recovery, as have the high interest rates. Even in those sectors in which the recovery has been most dramatic, capital is scarce in the eyes of management, because of the high interest rates and the need for new or modernized facilities to meet rising demand. Presented with the opportunity to convert to cogeneration, many managements seem to prefer to spend their money in more familiar ways—by building more plant to produce products for sale.

Third-party financing is nothing new, even for cogeneration. Some quite large cogeneration plants have come into being by that route in recent decades. It seems to be an unusually appropriate method for the present situation, however.

Cogeneration is highly capital intensive, so its development is somewhat dependent on the techniques available for financing. Cogeneration lends itself quite well to third-party financing. It produces large revenues or cost reductions; substantial benefits result from accelerated depreciation and the investment tax credit (which are fortunately still available in the final form of the 1984 Tax Reform Act); the nature of the assets permits a high degree of debt financing; and there is little technological risk. All of these advantages have caught the eye of the financial community in the past few years, and many firms are thus seeking out cogeneration projects to finance.

Implementation of an industrial cogeneration project requires the convergence and coordination of a number of factors: the industrial need for power and steam, the desire of industrial management to proceed, evaluation of the opportunity, design and engineering of the system, negotiation of contracts and securing of necessary approvals, construction management, and operating expertise—among others. It is not always easy to assemble all of these skills and properly coordinate them. Especially, it is not easy to find all of them based on real-life practical experience in the field. As a natural result, not all cogeneration projects in recent years have been or will turn out to be outstanding successes.

Recognition of this situation has led to the evolution of an essentially new industry—that of cogeneration development. These firms will design, build, and operate cogeneration plants, selling electricity and useful heat at a discount to a customer. In effect, they are acting as non-regulated utilities—able to supply energy cheaper to specific customers because they can take advantage of the higher efficiency of cogeneration.

We might think of the owning and operating of cogeneration facilities by third parties as analogous to the oil drilling business, and the cogeneration developer as analogous to the oil driller. An oil driller obtains a lease from a landowner and explores for oil on his property. If the well is a producer, the landowner receives his 1/8 royalty—even though he spent no money and took no risk of failure.

In a similar manner, a cogeneration developer "drills" for energy savings with cogeneration projects. Instead of oil, the resource is the energy user's demand for useful heat, which makes it possible for the developer to cogenerate electricity. In return for the right to use the energy user's heat demand to make cogeneration possible, the developer pays a royalty in the form of a discount on power and heat supplied to the customer.

Many things go into making an oil driller successful; there are also many factors necessary to make a cogeneration company successful. Technical expertise grounded in hands-on experience is one of the most important factors. The basic concepts of cogeneration are relatively simple, but the practice of cogeneration is complex—and difficult for those not soundly versed in the technology.

All the various aspects of combustion, steam generation, electric generation and distribution must be integrated effectively into a system design for a particular application. Detailed engineering must fit this system into the given location, as well as meeting many other important considerations. Control systems must be designed to meet the objectives of the project. Construction must be managed properly for greatest efficiency and to meet schedules. A wide variety of technical problems may surface during the start-up and must be dealt with promptly. When all is running smoothly, there is still the matter of good maintenance and operating organizations and procedures to be put into place—so that the savings projected will actually be achieved.

That is just on the technical side. There is much more to the story, if a company is to be successful. Cogeneration companies are no exception to this rule. There are probably more balls to keep in the air in this business than in many others. Many skills are required in addition to the technical, organizational and managerial expertise involved in the design, engineering, construction and operation of a cogeneration facility. Just to mention an obvious one—a thorough knowledge of federal, state and local regulations that may apply to a project is exceedingly important, and the resources to acquire all this essential information is an absolute necessity.

A less obvious area is insurance. One might think that a cooperative and skilled insurance agency could provide all the expertise required, but we have found that almost none of the large insurance companies quite understand what the cogeneration business really is. They do not yet have rates established for this kind of company, and it is necessary to work very closely with the agency, in order to get all of the needed coverage without being unduly penalized.

Two other aspects of the cogeneration business are virtually as important as technical expertise. The first is marketing. You must define the market segment you select as your target. You must become familiar with the priorities this market segment places on various kinds of activities, and where energy cost savings would rate on this priority scale. You must discover what kind of marketing approaches will work in this market, and which ones you should avoid. You must learn how the decision-making process typically operates in the organizations thought to be prospective customers.

Another key area is financing. A lot has been written on the various ways of financing cogeneration projects. But no book told us nearly as much about third-party financing of cogeneration projects as did the experience we gained in setting up our first limited partnership and selling the shares. One last comment on financing—for a small, new company on the scene, the most immediate problem is likely to be the financing of work under construction. And at this point, reputation, experience, and proven expertise in cogeneration of the key people in the new company become valuable assets.

Many companies are now in the cogeneration business, and many more will enter it as its attractiveness is discovered by more people. No doubt many of these firms will not really be qualified, and will fail. Unfortunately, this may give cogeneration a bad name for a time. Industrial companies interested in cogeneration and in third-party financing will be well advised to consider very carefully the qualifications of cogeneration developers who approach them. Contracts for cogeneration projects are relatively long-term, and potential industrial customers should make sure that they are involved with a cogeneration company with the skill and resources to be around for the long run.

The future for cogeneration looks bright. Some published projections indicate that by the year 2000, over 60 billion dollars will have been spent on cogeneration equipment and services. In that same year, it is expected that about 15 percent of the electricity produced in the United States will come from cogeneration.

Many new technologies will come to commercial realization in the 1980's and 1990's, expanding the market and improving the economics. Gas turbines with higher firing temperatures will increase the fuel savings. Fuel cells will allow electricity to be generated without the theoretical limitations of thermal cycles, and the waste heat can be used to create a highly efficient cogeneration system on a very small scale. Coal gasification, while not of itself a cogeneration technology, will permit the most efficient cogeneration systems to burn coal-derived fuel rather than oil or natural gas.

Cogeneration will be implemented intensively at first in regions with high electric rates—notably California and the Northeast. Other areas which are served by a utility with nuclear problems could also offer attractive cogeneration opportunities. As electric rates continue to escalate, along with other energy prices, cogeneration will spread across the entire United States.

The time is not far off, for many energy users, when it will no longer be truly economical to buy electricity generated in conventional central station power plants. For many, a move toward cogeneration will become necessary for survival in the highly competitive economic climate. The cogeneration industry stands ready to help them in making that move.

# CHAPTER 11
# How Well are PURPA's Cogeneration Incentives Functioning?

*Claire Wooster and Eric Thompson*

Numerous articles have been written on the regulatory and financial incentives offered private cogenerators in the Public Utility Regulatory Policy Act. Few, however, have attempted to chronicle in any detail how states have implemented the major provisions of PURPA or how industry has responded to these legislative incentives.

Perhaps the most positive developments for would-be cogenerators have been the two rulings by the Supreme Court upholding the federal legislative initiative. In June, 1982, the high court upheld the constitutionality of PURPA against the challenge of the state of Mississippi. Then in May, 1983, the same court reversed an appeal's court decision that would have vacated the Federal Energy Regulatory Commission's (FERC's) full-avoided cost rule and its blanket interconnection order.

As the result of these two decisions, it is now difficult to see how a further challenge to the implementation of the federal cogeneration incentives could be sustained.

Nevertheless, despite this resolution at the federal level, much uncertainty still remains for a firm considering an investment in cogeneration. As this chapter attempts to show, the multiplicity of approaches pursued by the different state public service commissions in fulfilling FERC's rules reveal the indefinite nature of the PURPA mandate at the state level and the continued uncertainty investors may face in attempting to evaluate the attractiveness of a cogeneration project.

A review of industrial response to date points out the importance of a state commission's active support of the substantive intent of PURPA for the realization of cogeneration projects.

## STATE PROCEEDINGS ON COGENERATION

The federal guidelines implementing PURPA left most decisions regarding the setting of rates for the purchase of power from qualifying facilities (QFs) and the sale of power to them in the hands of state regulatory agencies. State commissions were ordered to consider a number of factors in determining the costs a utility might avoid by the purchase of power from a qualifying facility.

Specifically, state commissions were required to set standard purchase rates for small cogenerators (those with a generating capacity of less than 100 kilowatts) and a method of purchase from larger qualifying facilities.

They also were required to set the rates, or means of sale, of supplementary, backup, maintenance and interruptible power by utilities to cogenerators. In addition, they had to set the conditions of interconnection—both in terms of the operational standards required and the costs to be borne by the cogenerator. Finally, they had to establish data filing requirements that would permit the would-be cogenerator to determine or verify the avoided costs likely to be paid by the utility.

In their initial orders, most states concentrated almost entirely on setting the methods for determining utility avoided costs and the rates for purchase of power by utilities from qualifying facilities (QFs). Only a few addressed in any detail the issues of sales of power to cogenerators or the establishment of guidelines for interconnection. The material that follows therefore reflects this emphasis.


## STATE APPROACHES
## TO SETTING PURCHASE RATES

Traditionally utility rates have been based on average or embedded costs. While a rising marginal cost curve for the electric utility industry has been noted by numerous commentators in recent years, few state commissions had studied marginal pricing concepts until mandated by PURPA to review rates based on these principles.

Thus, when required to set rates for self-generators based on the incremental costs avoided by the utility from such purchases, the procedures to carry out this mandate were entirely unfamiliar to most state commissions. Moreover, since all states were initiating procedures to establish these rates simultaneously, there were no model laws in other states to follow.

Theoretically the setting of avoided cost rates means that a utility's costs before any purchases of power from qualifying facilities are calculated, then its costs are recalculated after allowing for purchases from these facilities. The difference between these two figures is pro-rated to all facilities supplying power to the utility.

While the theory sounds simple, commissions immediately found that it hid numerous complexities. Not only do the numerous projections and hypotheses of the analytical model make commission verification of allowed costs difficult, but the various factors that FERC mandated commissions to consider when setting purchase rates implied that commissions might have to set a variety of purchase rates. Even FERC itself noted that "the translation of the principle of avoided capacity costs from theory into practice is an extremely difficult exercise . . ."[1]

While a utility might be purchasing 100 MW of cogenerated power, only 50 MW of this power might be flowing into the grid at the time of day when high demand requires the burning of expensive oil or the use of inefficient old plants. Moreover, of this 50 MW production, only 25 MW might be committed to the grid on long-term contract.

Thus, commissions had to grapple with such questions as: *How much more valuable is power available at peak hours than at off-peak hours? Is it necessary to know a utility's avoided costs and a self-generator's supply pattern on an hourly basis, or does this place an undue transactional burden on all concerned? Should the QFs supplying the 25 MW of long-term power be entitled to higher rates because the utility can count on their capacity in system planning? Or, since all QFs are not likely to cease production simultaneously, is it possible to assign some capacity value to noncontractual power in system planning?*

How commissions resolved these and a myriad of other issues greatly affected the final purchase rates and the resultant incentive for cogeneration.

Some states, either believing that self-generated power was likely to be an insignificant source of power for their utilities or that specifying detailed retail rates or methodologies was not an appropriate commission activity, did little more than restate the PURPA guide-lines in their orders.

They left the working out of appropriate methodologies and rates to utility discretion. By adopting this approach, commissions in these states sidestepped grappling with the thorny issues raised by avoided-cost ratemaking.

Other states, however, took the PURPA mandate seriously and wrote lengthy orders detailing the procedures to be followed in set-ting purchase rates. Some held extensive hearings on rates proposed for cogenerators by jurisdictional utilties. The methodologies dis-cussed below are ones found in a review of the orders and hearings of twenty-two states that chose, either through general rulemaking or the hearing process, to direct in some detail the procedures used by jurisdictional utilities in setting purchase rates.

### 1. The Energy Credit

The most common methodology prescribed by state commissions for determining avoided fuel costs has been termed the "incremental heat rate approach." This method involves estimating the heat rate of a certain increment of system load, and then multiplying this figure by the cost of the fuel required to produce these Btus.

Thus, at a time of day when system load only requires modern baseload plants fueled by coal or uranium, the energy cost of a kilo-watt of electricity is quite low. When inefficient older plants, peaking plants or oil-fired plants must come on line, however, then fuel costs rise considerably.

A variation on the incremental approach mandated by some com-missions involved the use of system lambda data and production cost models. Either historical or real time data on running costs of the utility on the margin were prescribed as the measure of avoided costs payable to the cogenerator. This method appeared to be popular with commissions that had sophisticated analytic models of their own.

If a production costing model was not used, then a suitable value had to be ascertained for the fuel cost applicable to the incremental heat rate. Since fuel costs are the major component of the energy credit, they are an important variable in establishing the final pur-chase rate.

The California PUC, for example, decided that since California utilities burned oil extensively, the appropriate measure of avoided fuel costs was the *estimated* (i.e., projected) cost of low sulfur fuel oil.

The Connecticut DPUC, on the other hand, although its utilities also burned oil extensively, mandated the use of the *historical, average* fossil fuel cost as determined over the previous 12-month period. Not surprisingly, Connecticut's approach resulted in a lower fuel credit than California's approach.

Some commissions decided that because of the difficulty in verifying utility production models or the expense involved for commissions and would-be cogenerators in independently duplicating these purchase figures, a better approach was to specify a generating unit for utilities to use in calculating avoided energy costs. While not as accurate in mirroring system costs, this approach made it easier for cogenerators to ascertain the energy credit they would receive for power produced. Thus, New Hampshire set a purchase rate based on the running costs of the most modern oil generating station in the state.

Another approach used by commissions wishing to provide a simpler and more easily verifiable purchase rate, was to require the use of the energy component of power purchased from a pool as a proxy for avoided costs. Both Iowa and New Jersey authorized the billing rate for energy purchased from a pool as the acceptable measure of avoided energy costs for utilities participating in pool arrangements. Many states also took this approach toward their nongenerating utilities, requiring them to pay cogenerators the price they paid for energy from a supplying utility.

Some states, however, noted that the price a utility paid a pool or another utility for power did not always reflect the running costs of the supplier. For example, a supplying utility's marginal running costs could be considerably higher than the embedded cost-based energy rate it was charging a nongenerating utility.

In power pools with central dispatch of power, the considerations could become even more complex because of the "split-the-savings" approach used by many utilities in buying and selling power among themselves. Purchase rates more or less favorable to cogenerators could be obtained according to which cost a commission determined to be a utility's actual avoided costs: the running costs of its supplier, the actual costs paid for power regardless of the supplier's costs, or some compromise thereof.

A number of states included other allowances or incentives in the energy credit. The most common allowance was for a reduction in line losses in the transmission of power. Since utilities have to generate more power than is ultimately consumed by the purchaser to account for losses occurring during transmission, some states gave cogenerators a credit to reflect their closer proximity to the ultimate point of consumption.

Another common allowance included variable running costs not already included in the primary calculation method. For example, if a commission did not require use of a production costing model, then inclusion of variable operation and maintenance costs might be required.

More idiosyncratic allowances included an "inflation factor," an "incentive for firm power," and a credit based on the benefit to the state and nation in reducing dependency on imported oil.

In the Spring of 1983, the energy component of standard purchase rates implemented under PURPA varied by more than a factor of eight. In Vermont a small qualifying facility could receive 9 cents per kilowatt-hour generated on peak, while in Indiana a similar facility could receive only 1.33 cents.

While a large part of this variation could be attributed directly to fuel costs (Indiana utilities burn coal almost exclusively, while Vermont utilities rely heavily on oil-fired generation in peak hours), some of the variation was also due to the methodologies chosen by commissions to calculate avoided energy costs and the additional costs or incentives they chose to include in this credit.

Commissions often had to choose between approaches that minimized costs to utilities, and thus, presumably, costs to the ratepayer, and one that offered more encouragement to cogenerators. Some commissions felt the extrinsic values of cogeneration, such as reducing the use of imported oil, increasing source diversification, or keeping power purchase within the local economy, more than offset the small additional burden marginal cost rate-setting might place on ratepayers. Others did not.

## 2. The Capacity Credit

Because FERC mandated payment of a capacity credit only if a utility had need for new capacity, all commissions did not include a determination of, or a method of determining, a capacity component in the standard rate. Alabama, Connecticut, Delaware, Rhode Island, and Vermont all pronounced their utilities as having excess capacity rendering unnecessary the need to develop a methodology for determining qualification for a capacity credit.

Other commissions, often appearing overwhelmed by the burden of trying to ascertain accurately the cost of avoided capacity, left the payment of a capacity credit to the discretion of the utility. Under this approach, would-be cogenerators were left to negotiate directly with the local utility for a capacity payment.

Some commissions appeared to be irritated at having to implement the payment of a credit that the FERC itself had admitted was difficult to determine. The Massachusetts DPU, while dutifully prescribing a procedure for calculation of a capacity credit for the standard rate, nevertheless called such a determination a "fool's errand." In spite of the difficulties involved, however, over half the states whose orders were reviewed made an attempt to prescribe a method for calculating the capacity credit.

The most common method prescribed for determining the capacity credit might be termed the "unit specified" approach. Under this approach the commission determined an appropriate generating unit for calculating the capacity payment.

Some chose to base a capacity credit on a gas turbine unit, since such units are frequently used to meet power needs in periods of peak demand. Others, however, chose to offer credits based on more expensive base-load power plants—particularly if the cogenerator was willing to enter a long-term contract.

A second method chosen by a number of commissions for its simplicity and ease of administration was that of pricing the capacity credit at the cost of capacity bought from a pool. Both Massachusetts and New Jersey determined that capacity-deficient utilities should pay cogenerators at the rate of the capacity deficiency charge they paid to their respective power pools.

A number of other commissions also took this approach towards nongenerating utilities, specifying that the capacity component of the wholesale rate of power purchased from another utility was an acceptable measure of avoided cost.

Yet another approach has been to set the capacity credit at the utility's carrying charge for prospective capacity. Thus, the Oklahoma Corporation Commission required an estimate of the annual carrying charge of the next generating unit to come on line, which was then divided by the number of hours this unit was expected to operate per year to determine the capacity payment per kilowatt-hour available to cogenerators.

A final, and more sophisticated, approach employed by a few commissions has been the "differential revenue requirement method." Under this approach, a utility determines its optimum capacity expansion plan without QFs. Then it reoptimizes the plan considering the input of QFs to the grid. The difference in capital costs between these two plans represents the avoided capacity costs available to QFs. Maine and Texas have both implemented variations of this approach.

While most correct theoretically, the differential revenue requirement method requires highly sophisticated planning models to implement. Many commissions simply to not have the computing or human resources either to carry out such an approach themselves or to verify utility computations using this approach.

Furthermore, implementation of this approach requires a large number of assumptions about future load requirements and the relative costs of various types of new generating capacity.

The wide variety of approaches employed by state commissions in setting capacity credits has made geographical location a crucial factor in obtaining attractive PURPA purchase rates. Even though it may employ system-wide planning, a given utility may be offering a variety of capacity credits as the result of varying mandates from different jurisdictional commissions.

The capacity credits payable by Utah Power, for example, at one point ranged from 26 to 35 to 47.9 mills per kilowatt-hour—depending on whether the cogenerator was located in Wyoming, Utah or Idaho respectively.

Then, too, because of the conditional nature of most capacity payments, the PURPA purchase rates available to many cogenerators are lower then the average cost-based retail rates. An energy credit alone, or an energy credit plus a minimal capacity credit, negates any benefit that simultaneous buy-and-sell was designed to provide.

Under these circumstances, then, the main incentive to cogenerate can only come from the avoidance of purchase of costly utility power. Industry would choose to meet internal load first and sell surplus power only to the local utility.

## DETERMINING ELIGIBILITY
## FOR THE PURCHASE RATE

While FERC had already set minimum ownership and efficiency standards for certification of cogenerators as qualifying facilities, state commissions frequently added criteria of their own for obtaining certain incremental payments on the energy credit or payment of the capacity credit.

In some cases they also relaxed FERC's criteria with respect to full avoided cost payments. In particular, a number of states decided that the standard purchase rates should be available to "old" as well as "new" cogenerators.

Similarly, FERC only required standard purchase rates to be made available to qualifying facilities of less than 100 kW capacity. A number of commissions determined, however, that the availability of standard rates to larger cogenerators would simplify their administrative burden and encourage good faith negotiation between utilities and cogenerators.

Both parties would then know the minimum purchase rate the cogenerator could legally demand. Accordingly, of the 36 states for which purchase rate information was available, 22 offered the standard rates to larger cogenerators—although one state (Pennsylvania) limited the rate to cogenerators with less than 500 kW capacity and six (Arkansas, Connecticut, Delaware, Maine, New Jersey, Utah) limited it to those with less than 1000 kW (1MW) capacity.

Most states offering a capacity credit made its payment contingent on the cogenerator fulfilling certain qualifying conditions. The most common condition specified was that the cogenerator deliver "firm" power. The cogenerator had to contract to supply a certain amount of capacity much as utilities might contract with one another to do.

Some states further specified a minimum contract length, arguing that unless the outside power supply could be counted on for a certain number of years, utilities could not integrate it into their system planning.

Some state commissions made the capacity credit contingent on the cogenerator's reliability. Because of utility fears that cogenerators would enter long-term contracts in order to receive capacity payments, but then would deliver only a small fraction of the contracted power, some commissions specified a minimum capacity factor that cogenerators had to meet in order to obtain a full capacity credit.

Other states avoided the necessity of determining a minimum capacity factor by requiring the cogenerator to contract for supplying a certain amount of energy, rather than a certain amount of capacity. Since only power produced in the hours of peak system demand helps to avoid the need for new capacity, these states usually also required peak-hour production for a full capacity payment.

## SETTING RATES FOR SALES TO COGENERATORS

As already noted, most of the states in their initial orders focused their attention on setting avoided cost purchase rates. Rulings on sales to cogenerators generally merely repeated the FERC guidelines.

Some commissions, most notably California and Idaho, noted that with simultaneous buy-and-sell, the QF is always buying electricity from the utility. Therefore, with an outage of QF equipment, there is only a reduction in utility purchases—utility sales to cogenerators remain constant.

Because outages of QF equipment do not affect sales by utilities, cogenerators should be treated just like other customers with similar load characteristics and pay similar rates. Futhermore, following this line of reasoning, rates for backup and supplementary power become irrelevant.

For those who chose to deliver surplus power only to the utility, however, the California commission did approve the use of a standby service rate schedule. In the case of Pacific Gas and Electric, the standby charge consists of a $5.00/month meter charge plus a $0.75/ kW month contract capacity charge.

If capacity is actually used, it is billed at the customer's applicable rate schedule. Total failure of the cogenerator's equipment, therefore, results in a charge only $5.00 greater than the paid by a normal firm service customer of similar load characteristics. Furthermore, the utility provides a total waiver of the standby charge for small customers (<300 kW capacity) using unconventional forms of generation.

PG&E's standby schedule offers a distinct contrast to that approved by the New York Public Service Commission for Consolidated Edison. The New York commission assumed that Con Ed's rate structure would induce sales of surplus power only to the utility. Therefore it approved an elaborate schedule of standby charges based on the assumptions of a 36% coincidence factor in the down time of cogeneration facilities during the peak Summer period.

Under Consolidated Edison's rate schedule, all standby customers are required to pay a meter charge of $6.00/month plus a contract capacity charge ranging from $2.01/kW/month for less than 10 kW capacity taken as low tension service on the secondary distribution network, to $4.22/kW/month for capacity in excess of 100 kW taken as high tension service on the primary distribution network.

There are penalty charges ranging as high as 24 times the monthly contract demand charge if the actual monthly demand exceeds the contract demand by more than 10 percent. In addition, the energy charge for large cogenerators (those whose contract demand exceeds 900 kW) during the summer on-peak billing period is 29.77 or 31.84 cents per kilowatt-hour, depending on whether high- or low-tension service is provided.

An obvious implication of these contrasting approaches is that rates paid by cogenerators to meet backup needs for maintenance and system emergencies may be a significant disincentive for new cogeneration projects. Where utility rates are high, energy costs avoided by the self-generation of power can, in themselves, produce sufficient incentive for new cogeneration projects. High standby rates, or the risks of severe penalties from failure of a new system to perform as planned, however, can severely reduce the return otherwise projected, and thus reduce the attractiveness of new projects.

Furthermore, few states have grappled with the thorny issues of backup rates. For example, should a utility assume that all QFs selling surplus power would require backup service at the same time, and thus need to be charged the full demand rates of regular service customers, or could an assumption of only a limited amount of "coincidence" be assumed so as to lower these rates? Until such issues are resolved more fully, high rates for sales to cogenerators may be as strong a disincentive to initiating cogeneration as low purchase rates.

## ESTABLISHING PROCEDURES
## AND COSTS FOR INTERCONNECTION

In their initial orders, many commissions merely repeated the FERC rule stating that incremental interconnection costs were to be borne by the QF, and suggested various payment schedules. Most left the determination of interconnection equipment necessary for utility safety and reliability to the purchasing utility.

Perhaps because of the activeness of Consolidated Edison in the court case opposing FERC's cogeneration rules, however, the New York commission issued an extensive order on interconnection. The commission required the QF to pay the costs incurred for delivery of power to the utility, which it further specified as: 1) the incremental metering charge; 2) a 9 percent carrying charge to cover taxes, operation and maintenance; 3) up front payment of first time interconnection costs; and 4) any costs associated with engineering and feasibility studies related specifically to parallel operations.

The detail and rigor of this order offers a distinct contrast to that of Rhode Island, where the commission only stipulated that costs associated with utility purchases should be borne by the qualifying facility, while those associated with sales should be borne by the utility. Moreover, the Rhode Island commission allowed the qualifying facility 5 years to reimburse the utility for first time interconnection costs. Again, the approach of the one commission appears to be much more encouraging of cogeneration than the approach of the other.

## INDUSTRIAL RESPONSE TO PURPA

Little attempt has been made to date to answer questions such as whether PURPA is actually encouraging more cogeneration, or whether some state approaches have been more encouraging than others. Perhaps the best place to begin looking for answers to these questions is in the FERC's quarterly report on cogeneration facility filings.

This report lists by region the name, address, facility type and rated capacity of those firms either notifying FERC of their qualifying status or requesting FERC certification in order to obtain the benefits of the PURPA legislation.

While this report clearly indicates the number and dispersion of firms that have been encouraged by the PURPA legislation to evaluate cogeneration, it probably does not indicate accurately the amount of cogeneration that actually will come on line in the next few years. A request for certification may be made at the initial planning stage of a project, but the project may then be altered or abandoned as the result of subsequent evaluation or unforeseen events.

A second problem with the filings list is that it has not been verified for definitional accuracy. Not only is there the possibility of error inherent in self-classification, but there is also the problem of dual qualification.

Two cogeneration filings in California, for example, specify "solar" as the cogeneration fuel. While one exceeds the size limitation for small power production facilities, the other presumably could have just as readily filed for certification as a small power producer.

Similarly, there are a number of firms with "Cogeneration" in their names that have applied for certification as small power production facilities. When a firm uses biomass as its primary fuel, it appears to qualify under either heading.

Therefore, firms presumably have applied for the type of certification that would yield the greatest tax benefits or other financial incentives for the project. Since FERC does not require notification of qualification under both definitions, only four filings note the possibility of this dual certification.

Table 11-1 is a listing of FERC's qualifying cogeneration filings by state as of the end of fiscal year 1983 (October, 1983). As can be seen, 204 cogenerators have filed for qualification of approximately 6,400 MW of new capacity. (This is equivalent to about six new baseload plants.) Over 40 percent, or 87 of these filings have come from California, while Florida and Texas have supplied another 15 percent (31 filings). In terms of rated capacity, however, Texas leads California with slightly under 40 percent of the total rated capacity. Just four states (Texas, California, Florida and Massachusetts), account for nearly 80 percent of the total capacity offered by new cogenera-

**TABLE 11-1. FERC Filings for Qualification as New Facilities***

| State | Number of Filings | Rated Capacity (kW) |
|---|---|---|
| Alabama | 1 | 37,400 |
| Alaska | 0 | 0 |
| Arizona | 0 | 0 |
| Arkansas | 1 | 375 |
| California | 87 | 1,577,112 |
| Colorado | 1 | 2,600 |
| Connecticut | 0 | 0 |
| Delaware | 2 | 70,139 |
| District of Columbia | 0 | 0 |
| Florida | 15 | 422,420 |
| Georgia | 2 | 76,600 |
| Hawaii | 1 | 120 |
| Idaho | 4 | 27,500 |
| Illinois | 4 | 3,078 |
| Indiana | 0 | 0 |
| Iowa | 1 | 10,000 |
| Kansas | 2 | 34,250 |
| Kentucky | 0 | 0 |
| Louisiana | 2 | 180,000 |
| Maine | 1 | 46,700 |
| Maryland | 1 | ? |
| Massachusetts | 3 | 583,400 |
| Michigan | 3 | 34,717 |
| Minnesota | 0 | 0 |
| Mississippi | 6 | 33,677 |
| Missouri | 1 | 80,000 |
| Montana | 1 | 12,000 |
| Nebraska | 0 | 0 |
| Nevada | 0 | 0 |
| New Hampshire | 1 | 1,800 |
| New Jersey | 3 | 53,300 |
| New Mexico | 0 | 0 |
| New York | 7 | 77,868 |
| North Carolina | 8 | 250,120 |
| North Dakota | 1 | 9,000 |
| Ohio | 1 | 16,500 |
| Oklahoma | 2 | 23,100 |
| Oregon | 0 | 0 |
| Pennsylvania | 5 | 155,315 |
| Rhode Island | 0 | 0 |
| South Carolina | 1 | 5,900 |
| South Dakota | 0 | 0 |
| Tennessee | 9 | 61,360 |
| Texas | 16 | 2,400,140 |
| Utah | 0 | 0 |
| Vermont | 0 | 0 |
| Virginia | 7 | 63,007 |
| Washington | 2 | 29,100 |
| West Virginia | 0 | 0 |
| Wisconsin | 0 | 0 |
| Wyoming | 2 | 17,000 |
| TOTALS | 204 | 6,395,598 |

*As of the end of fiscal year 1983.

tors. Thus, the PURPA legislation on cogeneration appears to have had minimal impact on the other 46 states.

The greater encouragement of cogeneration in these four states, as compared to the remaining 46, appears to be due to a number of factors. For a start, the rates offered to cogenerators appear to make simultaneous sale and purchase attractive. The PURPA purchase rates of Pacific Gas and Electric in California, for example, are higher than some of their retail sales rates. This also appears to be true of Tampa Electric in Florida, Boston Edison in Massachusetts, and Houston Light and Power in Texas.

Then, too, these states all offer standard rates to cogenerators with rated capacities of more than 100 kilowatts. As Table 11-2 shows, only 15 of the 204 qualifying facilities had a rated capacity of less than 100 kilowatts. Thus, unless state commissions explicitly offered the standard PURPA rates to larger installations, the remaining 190 facilities would have had to resort to individual negotiation. It would appear that perhaps the availability of a standard rate served to strengthen the cogenerator's bargaining position by acting as a legal minimum to which the cogenerator is entitled if negotiations falter.

### TABLE 11-2. Characteristics of New Facilities

*A. Facility Size*

| Capacity (kW) | # of Filings |
|---|---|
| $\leq$=100 | 15 |
| 101-1,000 | 48 |
| 1,001-10,000 | 65 |
| 10,001-100,000 | 64 |
| 100,001-1,000,000 | 11 |
| TOTAL | 203* |

*One filing failed to include any information on size.

*B. Fuel Type*

| Fuel | Number |
|---|---|
| B | 34 |
| C | 31 |
| C/B | 5 |
| FO | 7 |
| NG | 113 |
| SOL | 3 |
| WA | 10 |
| TOTAL | 203* |

*One filing did not provide information on fuel type.

Key:
B=biomass
C=coal
C/B=coal and biomass
FO=fuel oil, #2 or #6
NG-Natural gas
SOL=solar
WA=waste product (filer defined)

It would appear that perhaps the availability of a standard rate served to strengthen the cogenerator's bargaining position by acting as a legal minimum to which the cogenerator is entitled if negotiations falter.

In addition, all these states had rules and rates in place either by the FERC deadline of March 1981 or shortly thereafter. In contrast, three of the five states which still did not have final rules or rates as of March 1983 had no filings at all. Uncertainty over commission attitudes towards sale and purchase rates, therefore, may also have cast a cloud over the development of cogeneration in some states.

In fact, the perceived attitudes of state commissions towards cogeneration may be a major factor in encouraging its development. The California commission, for example, not only set favorable purchase rates, but also permitted cogenerators the same rate for natural gas as electric utilities pay.

In addition, it gave cogenerators priority over other industrial users in the case of gas shortfalls. As a further mark of its support of cogeneration, in 1979 the Commission reduced Pacific Gas and Electric's authorized return on equity by 0.2 percent because of its lack of effort to promote cogeneration. The return could only be regained if PG&E signed contracts for at least 600 MW of new cogeneration capacity in the next 2 years.

No other commission has assumed such an activist role in supporting cogeneration, nor has any other state come close to encouraging so much new cogeneration capacity. The natural gas incentives appear to have been particularly significant, since 48 of the 65 new filings in the state have listed natural gas as their fuel type.

California offers a particular contrast to New York. Both states are similarly populous and have similar fuel compositions when viewing the statewide utility system. But the New York commission has appeared to respond in a rather lukewarm fashion to cogeneration.

Because of the perceived slowness of the commission in setting PURPA purchase rates, the New York state legislature set its own minimum rate for all non-oil-fired facilities at 6 cents per kilowatt-hour. The commission did not issue a final order until May 1982, and then, while procedurally implementing the FERC's rules, the commission stated its desire to neither encourage nor discourage potential cogenerators in the Con Ed service area.

Subsequent to the order's issue, purchase rates still had to be put in place by all jurisdictional utilities. That New York state has had only one cogeneration filing, and that for a facility of only 100 kW, may be at least in part due to the lack of any strong encouragement of cogeneration by the state commission, as well as to the overt opposition of Consolidated Edison to cogenerators in its service area.

In states with PURPA purchase rates set below retail sales rates, cogeneration might still be encouraged if qualifying facilities had the option of net billing. Such an option has the effect of letting the QF avoid retail, embedded-cost payments for power used, while permitting sales of surplus power generated.

But a number of commissions either prohibited the net billing option or restricted it to small QFs. Under these conditions, the extra expense involved in selling surplus power only to the utility may well be the financial straw that deters the development of cogeneration.

As already noted, sales of surplus power require the establishment of reasonable backup or standby rates. Otherwise, the cogenerator is paying a full-service demand charge year round to cover service required during the 2 to 4 weeks of equipment outages necessitated by scheduled maintenance and emergency repair.

Commissions cannot be faulted in all states that show few or no filings of cogeneration capacity, however. The Idaho Commission, for example, perceiving Idaho Power's efforts to encourage cogeneration to be inadequate, not only threatened to set its rate of return at the "lower end of a range found to be reasonable," but also said,

> Failure to exhaust all power supplies available from cogeneration and small power production shall be grounds for rejection of applications for certificates of convenience and necessity regarding construction of conventional thermal units or for the issuance of securities to finance such units.

In spite of this activist stance, there have been only two cogeneration filings in Idaho.

Similarly, Vermont has had no cogeneration filings although it has offered one of the highest standard purchase rates in the country— 9.0 cents per kilowatt-hour at on-peak times and 6.6 cents off-peak. The obvious inference from the Vermont and the Idaho cases is that a certain level or type of industrial activity must already be present in a state for even strong PURPA incentives to have any effect. Such incentives are not enough to bring new industry into an area, although they may be sufficient to alter the energy consumption patterns or technologies of established industries.

Beyond the lack of an industrial base able to avail itself of the PURPA incentives, development almost everywhere has been discouraged by the uncertainties associated with PURPA itself. The American Paper Institute, for example, noted the PURPA could not become a factor in plant planning until FERC issued its rules.

Then many industries still felt they had to wait until their state commission implemented these rules. In the middle of the implementation process, however, the lower courts threw into doubt the constitutionality of PURPA and the validity of FERC's full avoided cost and interconnection rules.

Additional uncertainty has been created by the unpredictability of fuel prices. In the past year oil prices have declined considerably, resulting in a decline in the energy credit payment for those cogenerators who are paid on a real time or quarterly update basis.

The American Paper Institute has concluded that few businessmen will be willing to make the incremental investment PURPA requires faced with these uncertainties. "What is again missing as it was prior to PURPA, is a price signal that will be reasonably durable and certain."[2]

As the result, cogeneration generally only appears to be attractive to industry where at least one of three conditions prevail: 1) the company already is a cogenerator, so has existing expertise on its staff for making the most of the PURPA incentives; 2) a state commission or utility is actively encouraging potential cogenerators; or 3) the firm has an ancillary benefit to gain from cogeneration.

In the first category are firms such as Dow Chemical and a number of paper companies that were traditionally cogenerators. Their previous experience has placed them in the forefront of the return to cogeneration.

California is the outstanding example of a state whose regulatory policy has actively encouraged cogeneration. By removing potential barriers to cogeneration in natural gas regulation, offering accelerated depreciation on new cogeneration equipment, as well as by instituting strong PURPA incentives, it has managed to encourage almost as many filings as the rest of the states put together.

The companies responding to the third condition stated above are perhaps the most interesting. In this class are a number of firms with a waste product that is combustible. Faced with increasing costs for disposal of this product due to environmental regulation, these companies have decided to burn the product instead to meet thermal and electrical needs.

One of the most publicized of these ventures has been the cogeneration plant of the Diamond Walnut Growers cooperative in Stockton, California. By burning walnut shells, this group expects to earn $1 million a year on electricity sales and natural gas savings. Even in states where the price for cogenerated power is not favorable, however, the cogeneration option may still appear attractive if there are substantial waste disposal costs to be saved by its adoption.

## ASSESSING THE FUTURE

While there are still many factors inhibiting its development, the decline of cogeneration appears to have been reversed. Media reports, as well as the FERC filings, have indicated a resurgence of interest in this old technology. What is not yet clear, however, is the strength of this reversal. The authors are aware, for example, of two large projects listed in the FERC filings that may not materialize because of failure to obtain a purchase rate for the electricity produced to make the project economically feasible.

It is entirely possible that a comprehensive survey would reveal that no more than half of the capacity projected by these filings is actually being developed. A systematic follow-up of the FERC filings might not only give a better estimate of the amount of cogeneration coming on line, but also indicate the major stumbling blocks to project implementation.

On the basis of the filings themselves, it seems safe to conclude that while the PURPA legislation in itself has not given a strong boost to cogeneration, when coupled with a comprehensive state regulatory policy to encourage cogeneration the results can be quite dramatic. The large number of filings induced in California show that a coherent regulatory policy designed to remove a wide variety of institutional barriers can be an effective means of encouraging the growth of a technology.

The flexibility afforded states by FERC in implementing PURPA, however, appears to have resulted in wide variation in the degree of encouragement offered cogeneration. By leaving the determination of energy and capacity credits to jurisdictional utilities, rather than prescribing detailed methods themselves, many commissions have essentially permitted utilities to offer cogenerators whatever rates they see fit.

The lack of standard rates for facilities about 100 kW capacity, resulting in the need to engage in case-by-case negotiation with the utility, appears to have discouraged filings in a number of states. Industrial representatives have testified repeatedly of the problems of negotiating directly with many utilities. Even though utilities may be required to interconnect, unwilling utilities can effectively discourage development of private projects by offering low avoided cost purchase rates.

While the federal court challenges to PURPA should now be over, reports in *Energy User News* on cogeneration rate-setting under PURPA note the challenge of the final orders of Florida, Idaho, Kansas, Montana, New York and Pennsylvania in state courts by one or more jurisdictional utilities. Since all these states wrote fairly comprehensive orders detailing methods to determine PURPA rates, it appears that disgruntled utilities have adopted a new venue from which to oppose the avoided cost rule. Ross D. Ain, an attorney handling cogeneration cases, has estimated that it could take 5 to 10 years to resolve PURPA implementation firmly at the state level. If this estimate should prove to be correct, then legal uncertainties may have a dampening effect on the development of cogeneration projects in certain areas of the country for some time to come.

Therefore, between the reluctance of many utilities to negotiate what are perceived to be fair purchase rates by potential cogenerators and the litigious stance of some of the more stringently regulated utilities, it appears that industrial concerns considering cogeneration will still face difficulties in initiating projects for some time to come. In particular, these conditions will discourage firms that have little use for the electrical output themselves, but would wish to sell virtually all electricity generated to the utility at a price that produces an acceptable return on the incremental investment, projects are likely to be either terminated or mis-sized.

Under these circumstances, it appears that cogeneration would receive more encouragement if FERC both mandated standard purchase rates for larger facilities and set more explicit guidelines for calculating purchase rates. In particular, a more uniform approach to offering capacity credits would be useful.

Since the Commission acknowledged that setting capacity payments was indeed a difficult task, it seems reasonable to give states more guidance as to when such rates should be offered and what should be included in them. Investors would then have greater certainty in evaluating cogeneration projects.

While many industries oppose amending PURPA so that utilities might have majority ownership of qualifying facilities, such an amendment might encourage more cogeneration if private projects were not obstructed as the result. By requiring utilities to set standard rates applicable to all qualifying facilities, including their own cogeneration subsidiaries, utilities might then be more motivated to set true avoided cost rates that would benefit all cogenerators.

The history of the PURPA legislation to date shows how difficult entrenched legal barriers to a technology are to remove. Currently cogeneration appears to be making gains where state commissions have chosen to take the substantive mandate to encourage cogeneration seriously, rather than merely to implement PURPA procedurally.

As long as legal skirmishes continue at the state level, however, or buy-and-sell rates remain unattractive, many businesses will remain unwilling to commit funds to such projects. Therefore, while the federal legislative effort to encourage cogeneration has made notable gains, the revival is yet a tenuous one.

## FOOTNOTES

[1] 45 Federal Register 12214 (1980) at 12226.

[2] *Hearing Before the Subcommittee on Energy Regulation of the Committee on Energy and Natural Resources on S. 1885-Part I,* United States Senate, 97 Cong., 2nd Sess., April 19, 1982, at 333.

# CHAPTER 12

# Are Power Exchange Services Available to Cogenerators?

*W.E. Brand*

One of the most popular pieces of the 1978 Public Utility Regulatory Policies Act with industrial electric power users was Section 210, 16 USC § 824a–3. That provision was designed to encourage the development of cogeneration by requiring utilities to deal with cogenerators—to interconnect with them and to buy and sell power over the interconnection.

Objections of the utility industry to two aspects of the Federal Energy Regulatory Commission's rules implementing the statute were resolved by the Supreme Court on May 15, 1983, in the case of *American Paper Institute v. American Electric Power Service Company*, 103 S.Ct. 1921 (1983). These are (1) the use of the full avoided cost rather than some part of it as the cogenerator and (2) the requirement of interconnection by the utility without the inevitability of a costly full factual hearing in each case.

Under the recent economics of a power supply with rapidly increasing long run marginal costs, it has frequently been best for industrial cogenerators which utilize firm power in their manufacturing processes to buy the firm power they need from a utility at the average cost based rate of the utility and sell to the utility surplus energy (and capacity, if any) and the utility's full avoided or decremental cost.

These economics may be changing. Old, low cost gas contracts are expiring. New nuclear and coal base load units with high long run marginal cost are being plugged into the rate base. Both these factors are tending to raise the average cost rates.

Also, those rates include the utility's transmission and distribution costs whereas the cogeneration cost is a busbar cost. The utility's average cost rates may also be tilted in favor of the residential customer class—particularly if the state commissioners are elected.

The costs of cogeneration, on the other hand, are what they are. As a result, after study, an industrial cogenerator may prefer to keep its secondary energy resource and buy from the local utility those "power exchange services" which, when combined with that resource, produce service equivalent to "firm power" service and usable by the industrial in its manufacturing process, instead of buying "firm power" at average cost and selling its surplus energy.

For those who are not familiar with the concepts of "firm power" and "power exchange services," it may be helpful to explain these before reviewing that part of FERC's regulations which may provide industrials the opportunity to obtain such power exchange services.

Firm power is not available from any single thermal generator because such generators have a relatively high incidence of forced outage. It is not unusual for a deisel engine or a steam-turbine (and its associated boiler) to have rates of forced outage in the order of 2%-15% and perhaps even higher in the first one or two years of installation before the unit matures.

With even the very lowest rate of forced outage for thermal generating units, for example, 2% of the time, power would be unavailable 2 days out of 100 or about 7 days of every year. No residential customer would want to rely on that kind of power supply for home use and only a few industries which are able to use "interruptible power" would care to rely on power supply with that incidence of forced outage.

To enable it to sell power with greater reliability, a utility will interconnect two or more generating units to a single transmission system and supply energy from the pooled resources. A common rule of thumb for "firm power" is to be able to supply the full load at time of annual peak load despite the occurrence of a severe forced outage contingency. The contingency usually selected as the standard is the loss of the largest generating unit on the system at the time of the system peak load.

Figure 12-1 illustrates three ways in which a hypothetical utility could design a system to market 10,000 kilowatts (10MW) of firm power under the planning constraint that it must be able to withstand loss of the largest unit at the time of maximum load. Since

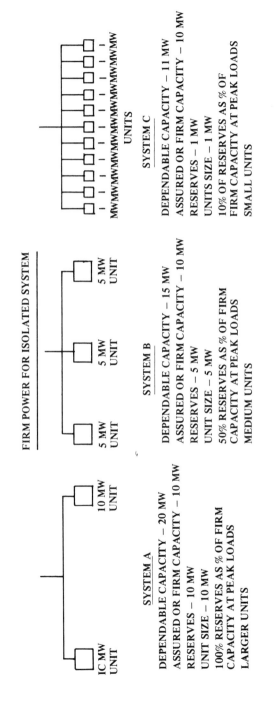

FIGURE 12-1.

FIRM POWER FOR ISOLATED SYSTEM

**SYSTEM A**

DEPENDABLE CAPACITY – 20 MW
ASSURED OR FIRM CAPACITY – 10 MW
RESERVES – 10 MW
UNIT SIZE – 10 MW
100% RESERVES AS % OF FIRM
CAPACITY AT PEAK LOADS
LARGER UNITS

**SYSTEM B**

DEPENDABLE CAPACITY – 15 MW
ASSURED OR FIRM CAPACITY – 10 MW
RESERVES – 5 MW
UNIT SIZE – 5 MW
50% RESERVES AS % OF FIRM
CAPACITY AT PEAK LOADS
MEDIUM UNITS

**SYSTEM C**

DEPENDABLE CAPACITY – 11 MW
ASSURED OR FIRM CAPACITY – 10 MW
RESERVES – 1 MW
UNITS SIZE – 1 MW
10% OF RESERVES AS % OF
FIRM CAPACITY AT PEAK LOADS
SMALL UNITS

PRINCIPLE NO. 1 – SYSTEM PLANNING FOR ISOLATED SYSTEM IS A COMPROMISE BETWEEN ECONOMIES AVAILABLE
FROM INCREASING UNIT SIZE AND DISECONOMIES OF LARGE FIXED COSTS OF RESERVES BECAUSE
OF IDLE RESERVE CAPACITY.

there are substantial economies of scale in larger individual unit sizes, it turns out that in planning electric systems to furnish firm power, system planning is a compromise between the economies available from increasing unit size and the diseconomies of the large fixed costs of reserves because of large amounts of idle reserve capacity.

"Reserve sharing" is a way to minimize the cost of idle reserves and still use large unit sizes. Figure 12-2 illustrates how two or more utilities can reduce reserves.

Figure 12-2 shows how interconnection reduces the percentage of reserves required to maintain the same assurance of firm power.

Figure 12-3 shows how reserve sharing works. Although each system is obligated to supply any surplus "emergency energy" to the other, by itself such emergency energy supply is not firm. When combined with the receiving system's resources, however, it becomes firm. Why? Because the probability of simultaneous but random outages on both systems is very low. This gives the receiving system assurance that it will have available, *either* from its own reserve, *or* over its interconnection with others, sufficient capacity to meet its own peak at all times except for an extremely small fraction of the time that is tolerable even for "firm power."

But because the emergency power is not firm, the supplying utility imposes no demand charge. The only requirement is that both systems maintain the same minimum reserve percentage. It is customary that the emergency energy sold is at the supplier's hourly incremental cost. Since the emergencies are usually of short duration, no great penalty is involved.

Why would a utility be reluctant to enter into a reserve sharing arrangement with an industrial? For the same reason many are reluctant to enter into such arrangements with small municipals and cooperatives. This is because the small system gains much more from the transaction than the larger one, assuming the transfer terms are fair. See Figure 12-4. This makes the smaller system or plan of industrial cogeneration more competitive with the rate of the larger utility.

The FERC's authority to order a plan of reserve sharing between small and large utilities, based on each of them maintaining the same percentage of reserves, was upheld by the Supreme Court in *Gainesville Utilities Department v. Florida Power Corporation*, 402 U.S. 515 (1971). In addition, the Nuclear Regulatory Commission had

**FIGURE 12-2.**

RESERVE SHARING
HOW IT AFFECTS COSTS OF FIRM POWER

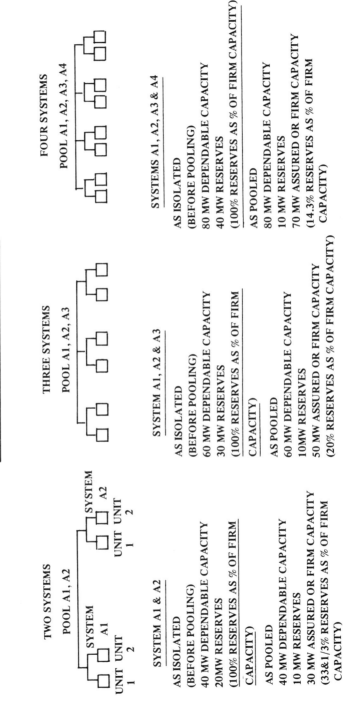

**TWO SYSTEMS**
POOL A1, A2

**SYSTEM A1 & A2**

AS ISOLATED
(BEFORE POOLING)
40 MW DEPENDABLE CAPACITY
20MW RESERVES
(100% RESERVES AS % OF FIRM CAPACITY)

AS POOLED
40 MW DEPENDABLE CAPACITY
10 MW RESERVES
30 MW ASSURED OR FIRM CAPACITY
(33&1/3% RESERVES AS % OF FIRM CAPACITY)

**THREE SYSTEMS**
POOL A1, A2, A3

**SYSTEM A1, A2 & A3**

AS ISOLATED
(BEFORE POOLING)
60 MW DEPENDABLE CAPACITY
30 MW RESERVES
(100% RESERVES AS % OF FIRM CAPACITY)

AS POOLED
60 MW DEPENDABLE CAPACITY
10MW RESERVES
50 MW ASSURED OR FIRM CAPACITY
(20% RESERVES AS % OF FIRM CAPACITY)

**FOUR SYSTEMS**
POOL A1, A2, A3, A4

**SYSTEMS A1, A2, A3 & A4**

AS ISOLATED
(BEFORE POOLING)
80 MW DEPENDABLE CAPACITY
40 MW RESERVES
(100% RESERVES AS % OF FIRM CAPACITY)

AS POOLED
80 MW DEPENDABLE CAPACITY
10 MW RESERVES
70 MW ASSURED OR FIRM CAPACITY
(14.3% RESERVES AS % OF FIRM CAPACITY)

**FIGURE 12-3.**

GENERAL REQUIREMENTS FOR
RESERVE SHARING ARRANGEMENTS

1. PHYSICAL INTERCONNECTION OF ADEQUATE CAPACITY
2. CONTRACT TO SUPPLY EMERGENCY POWER "IF-AND-WHEN-AVAILABLE." THIS MAKES POWER OF INTERCONNECTED SYSTEM FIRM SINCE PROBABILITY OF $A_1$ & $A_2$ HAVING A SIMULTANEOUS OUTAGE IS VERY LOW.
3. ASSUMES:

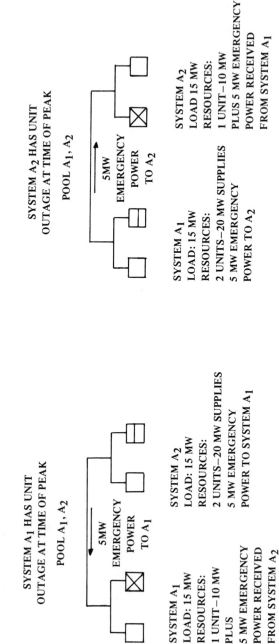

SYSTEM $A_1$ HAS UNIT
OUTAGE AT TIME OF PEAK

POOL $A_1$, $A_2$

5MW
EMERGENCY
POWER
TO $A_1$

SYSTEM $A_1$
LOAD: 15 MW
RESOURCES:
1 UNIT—10 MW
PLUS
5 MW EMERGENCY
POWER RECEIVED
FROM SYSTEM $A_2$

SYSTEM $A_2$
LOAD: 15 MW
RESOURCES:
2 UNITS—20 MW SUPPLIES
5 MW EMERGENCY
POWER TO SYSTEM $A_1$

SYSTEM $A_2$ HAS UNIT
OUTAGE AT TIME OF PEAK

POOL $A_1$, $A_2$

5MW
EMERGENCY
POWER
TO $A_2$

SYSTEM $A_1$
LOAD: 15 MW
RESOURCES:
2 UNITS—20 MW SUPPLIES
5 MW EMERGENCY
POWER TO $A_2$

SYSTEM $A_2$
LOAD 15 MW
RESOURCES:
1 UNIT—10 MW
PLUS 5 MW EMERGENCY
POWER RECEIVED
FROM SYSTEM $A_1$

**FIGURE 12-4.**

RESERVE SHARING
UNEQUAL SIZE SYSTEMS

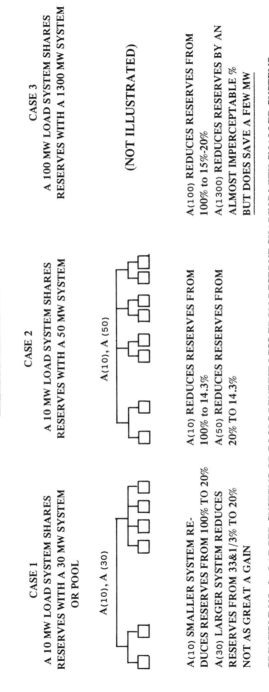

**CASE 1**
A 10 MW LOAD SYSTEM SHARES RESERVES WITH A 30 MW SYSTEM OR POOL

A(10), A (30)

A(10) SMALLER SYSTEM RE-DUCES RESERVES FROM 100% TO 20%
A(30) LARGER SYSTEM REDUCES RESERVES FROM 33&1/3% TO 20% NOT AS GREAT A GAIN

**CASE 2**
A 10 MW LOAD SYSTEM SHARES RESERVES WITH A 50 MW SYSTEM

A(10), A (50)

A(10) REDUCES RESERVES FROM 100% to 14.3%
A(50) REDUCES RESERVES FROM 20% TO 14.3%

**CASE 3**
A 100 MW LOAD SYSTEM SHARES RESERVES WITH A 1300 MW SYSTEM

(NOT ILLUSTRATED)

A(100) REDUCES RESERVES FROM 100% to 15%-20%
A(1300) REDUCES RESERVES BY AN ALMOST IMPERCEPTABLE % BUT DOES SAVE A FEW MW

PRINCIPLE NO. 2 - LARGER SYSTEMS OR POOLS BENEFIT LESS FROM RESERVE SHARING WITH SMALLER SYSTEMS ON COMPETITIVE TERMS THAN DO SMALLER SYSTEMS.

held that a utility's refusal to engage in reserve sharing on "*Gaines-ville* terms" was inconsistent with a utility's obligations under anti-trust principles, as applied under Section 105c of the Atomic Energy Act of 1954, as amended, 42 USC § 2135(c), to utilities seeking licenses to construct nuclear generating units. *Consumer Power Company*, 6 NRC 892, 1067-78 (1977).

Section 292.305(b) of FERC's regulations under PURPA implies a similar obligation under the cogeneration statute. II FERC Statutes and Regulations ¶ 25,135. It states:

"(b) *Additional services to be provided to qualifying facilities.* (1) Upon request of a qualifying facility, each electric utility shall provide:
(i)    Supplementary power;
(ii)   Back-up power;
(iii)  Maintenance power; and
(iv)   Interruptible power.
(2)    The State regulatory authority (with respect to any electric utility over which it has ratemaking authority) and the Commission (with respect to any nonregulated electric utility) may waive any require-ment of paragraph (b) (1) of this section if, after notice in the area served by the electric utility and after opportunity for public com-ment, the electric utility demonstrates and the State regulatory authority or the Commission, as the case may be, finds that com-pliance with such requirement will:

(i)    Impair the electric utililty's ability to render adequate service to its customers; or

(ii)   Place an undue burden on the electric utility.

(c) *Rates for sales of back-up and maintenance power.* The rates for sales of back-up power or maintenance power:
(1) shall not be based upon an assumption (unless supported by factual data) that forced outages or other reductions in electric output by all qualifying facilities on an electric utility's system will occur simultane-ously, or during the system peak, or both; and
(2) shall take into account the extent to which scheduled outages of the qualifying facilities can be usefully coordinated with scheduled outages of the utility's facilities."

Power system planners evaluating the economic feasibility of co-generation are overlooking an important alternative if they fail to consider the availability of these kinds of power exchange services in evaluating specific projects.

# CHAPTER 13
# Environmental Assessment of Gas-Fueled Cogeneration

*F.R. Kurzynske, C.R. Cummings*
*E.W. Hanson*

Interest in natural-gas-fired cogeneration systems is on the rise due to a combination of rapidly escalating energy prices, cogeneration's inherently high efficiency and a multitude of recent pro-cogeneration legislation at the federal and state levels. Presently, cogeneration is viewed by many as a unique and economically attractive business venture as well as a cost and energy-saving opportunity for gas customers. An understanding of environmental issues associated with cogeneration should enable better cogeneration investment decisions to be made.

## ENVIRONMENTAL REGULATORY REQUIREMENTS

Developers/owners of a cogeneration project may need to address, dependent upon project size and site, a mixture of federal, state and local environmental regulations. Federal and state agencies charged with enforcing these regulations have been surveyed for specific regulations affecting proposed cogeneration projects.

The requirements identified fall into four categories: 1) state environmental policy acts; 2) state energy facility siting acts; 3) federal National Ambient Air Quality Standards; and 4) federal New-Source Performance Standards.

Twenty-two states have enacted environmental policy acts, and twenty-five states have passed energy-facility siting acts, both of which may apply to cogeneration projects. The size of the proposed facility usually is the determining factor. A number of states require an environmental review for stationary emission sources as small as one million Btu/hr input, or about 100 kW. Table 13-1 shows the states that have legislated either or both of these acts.

TABLE 13-1. State-Level Environmental Policy Acts and Energy-Facility Siting Laws.

| State | Environmental Policy Act, Date and Type[a] | Energy-Facility Siting Law | | |
|---|---|---|---|---|
| | | Date If Enacted | Size of Plant Covered[b] | Comments on Applicability to Industrial Plants |
| Arizona | Yes (1971) SL | Yes (1971) | ≥100 MW | Any thermal, electric, nuclear or hydroelectric plant |
| Arkansas | No | Yes (1973) | ≥50 MW | Any electric generating plant and associated fuel supply |
| California | Yes (1970) SR | Yes (1974) | ≥50 MW | Any electric generating facility |
| Connecticut | Yes (1973) SR | Yes (1974) | All | — |
| Delaware | Yes (1973) SL | No | — | — |
| Florida | No | Yes (1973) | All | Any steam or solar electric generating facility |
| Georgia | Yes (1972) SL | No | — | — |
| Hawaii | Yes (1974) SR | No | — | — |
| Indiana | Yes (1972) SR | No | — | — |
| Iowa | No | Yes (1976) | ≥100 MW | Any electric generating plant |
| Kansas | No | Yes (1976) | All | Any plant used to produce electric power |
| Kentucky | No | Yes (1974) | All | Any facility for generating electricity for public consumption |
| Maryland | Yes (1972) SR | Yes (1971) | All | — |
| Massachusetts | Yes (1972) SR | Yes (1973) | ≥100 MW | Any bulk electricity generating unit |
| Michigan | Yes (1971) AO | No | — | — |
| Minnesota | Yes (1973) SR | Yes (1973) | ≥50 MW | Any electric power generating equipment |
| Montana | Yes (1973) SR | Yes (1973) | ≥50 MW | Also applies to plants producing liquid or gaseous fuels |
| Nebraska | Yes (1973) AO | No | — | — |
| Nevada | Yes (1971) AO | Yes (1971) | All | Any electric generating facility (electric utilities) |

| State | | | | |
|---|---|---|---|---|
| New Hampshire | No | Yes (1971) | ≥50 MW | Any electric generating station or equipment |
| New Jersey | Yes (1973) AO | Yes (1973) | All | Any industrial plant |
| New Mexico | No | Yes (1971) | ≥300 MW | — |
| New York | Yes (1975) SR | Yes (1972) | ≥50 MW | Any steam electric generating facility |
| North Carolina | Yes (1971) SR | No | — | — |
| North Dakota | No | Yes (1975) | ≥50 MW | Any plant capable of producing electricity |
| Ohio | No | Yes (1972) | ≥50 MW | Any electric generating plant |
| Oregon | No | Yes (1972) | ≥25 MW | Any electric power generating plant |
| South Carolina | No | Yes (1971) | ≥75 MW | Any electric generating plant |
| South Dakota | Yes (1974) SR | No | — | — |
| Texas | Yes (1972) AO | No | — | — |
| Vermont | No | Yes (1975) | All | — |
| Virginia | Yes (1973) SR | No | — | — |
| Washington | Yes (1971) SR | Yes (1970) | ≥250 MWc | Any electric generating facility for distributing electricity for electric utilities |
| Wisconsin | Yes (1971) SR | Yes (1975) | ≥12 MW | Any bulk electric generating facility |
| Wyoming | No | Yes (1975) | ≥100 MW | Any energy generating facility |

[a] Environmental policy acts are of three principal types: comprehensive statutory requirements (SR), including actual legislation and comprehensive requirements as part of the law; comprehensive or administrative orders (AO), including administrative orders, no actual legislation, and comprehensive requirements as part of the order; special or limited environmental impact statement requirements (SL), including limited applications (e.g., coastline projects, energy production facilities).

[b] The size of plant covered is typically vague with respect to specifying thermal or electrical energy capacity. Most laws apply to electric generating facilities but do not state specifically that the size limitations refer to electric generating capacity.

Source: Icerman, Larry, *et al.*, *Environmental Constraints on Implementation of Industrial Cogeneration Facilities*, 1980.

The Environmental Protection Agency has established primary (health) and secondary (welfare) National Ambient Air Quality Standards (NAAQS) for total suspended particulate, ozone, carbon monoxide, sulfur dioxide, nitrogen dioxide and lead emissions. The standards set maximum ambient levels for these pollutants. If the local air exceeds the level specified, the area is considered to be a non-attainment area for that pollutant. To prevent deterioration of air quality in attainment areas, and to improve air quality in non-attainment areas, the EPA has set a number of permitting requirements applicable for new pollution sources, such as cogenerators.

A proposed cogeneration facility would be required at a minimum to satisfy the New-Source Performance Standards (NSPS) presented in Table 13-2. In addition, a cogenerator may be subject to Best Available Control Technology (BACT) or Lowest Achievable Emission Rate (LAER), depending on its location (attainment or non-attainment area) and its estimated net potential emissions.

If the facility is sited in an attainment area and it emits 100 tons or more annually of any of the regulated pollutants, the cogenerator is subject to Prevention of Significant Deterioration (PSD) Regulations. The PSD review requires the facility to demonstrate use of BACT, which is performed on a case-by-case basis, but in no instance is less than NSPS.

If the facility is sited in a non-attainment area and emits 100 tons or more of the pollutant for which the area is designated "non-attainment," it will be subject to a New Source Review (NSR). The NSR requires that the facility:

- utilize lowest achievable emission rate (LAER) controls;
- certify that all existing sources in the state are in compliance or are following an approved schedule for compliance;
- provide emission reduction capabilities (off-sets) from existing sources in the area; and
- demonstrate that the offsets will produce a positive net air quality benefit.

Clearly, it is advantageous to avoid an NSR or PSD review. This can be accomplished by sizing the facility small enough so that the net new emissions do not exceed the 100-ton/year limit. Table 13-3 gives the approximate sizes of new gas-fueled cogenerators that will

TABLE 13-2.
New-Source Performance Standards for Gas-Fueled Cogenerators

| Emission Source | Pollutant | | | |
|---|---|---|---|---|
| | NOx | SO$_2$ | Particulates | Visual |
| Fossil Fuel Fired Steam Generators[a] | | | | |
| ≤250 MMBtu/hr input | — | — | — | — |
| >250 MMBtu/hr input | 0.2[b] | — | 0.1[b] | 20% Opacity |
| Gas Turbines:[k] | | | | |
| <3 MW[c] | — | — | — | — |
| ≥3 MW to 30 MW[d] | 150 ppm[e,f] | 150 ppm[e,g] | — | — |
| ≥30 MW (Utility)[d] | 75 ppm[e,f] | — | — | — |
| ≥30 MW (Industrial)[d] | — | — | — | — |
| Emergency | — | — | — | — |
| Reciprocating Engines:[h] | | | | |
| <560 CID/Cyl. | — | — | — | — |
| ≥560 CID/Cyl. | 600[i,j] | — | — | — |

[a] 40 CFR 60.42, 43, 44

[b] Pounds per MMBtu of fuel

[c] 3 MW = 10.72 gigajoules

[d] 30 MW - 107.2 gigajoules

[e] At fifteen percent oxygen, dry basis

[f] Corrected for heat rate at specific load conditions and nitrogen content of the fuel

[g] Or 0.8% sulfur limit in the fuel

[h] Federal Register, Vol. 44, July 23, 1979, No. 142 (Proposed)

[i] Adjusted upward for thermal efficiencies greater than 35%

[j] Measured emissions adjusted to standard atmospheric conditions of 29.92 inches Hg, 85°F, at 75 grains moisture/lb of dry air, and fifteen percent oxygen on a dry basis

[k] 40 CFR 60.332, 333

not require a Federal review. If there are offsets available, such as a boiler that will be displaced, the cogenerator could be larger.

## EMISSIONS PERFORMANCE

The combustion of natural-gas is an inherently clean process, producing almost no particulates or sulfur dioxide and very little carbon monoxide or hydrocarbons. The only pollutant produced

**TABLE 13-3.**
**Approximate Maximum Size of Gas-Fueled Cogeneration**
**Systems to Avoid Federal NSR/PSD Review.**

| Cogeneration System | Electrical Output (MW) |
|---|---|
| Reciprocating engines[a] | |
| Proposed NSPS NOx Limit | 2.5 |
| Gas turbines[b] | |
| Gas-fired NOx NSPS Limit | 11.5-17.0 |
| Steam turbines[c] | |
| Gas-fired NOx Limit | 1.6-2.2 |

[a] Total plant efficiency 30%
[b] Total plant efficiency 20-30%
[c] Total plant efficiency 10-15%
Source: Office of Technology Assessment

in quantity is nitrogen oxides (NOx). Therefore, the emissions performance of gas-fueled systems focuses on NOx only.

NOx emissions data for both uncontrolled and controlled systems were obtained from manufacturers, agencies and the literature. Typical NOx emission factors are presented in Table 13-4. The data in Table 13-4 are for comparison purposes only. Actual emission factors of cogenerators may vary significantly depending on equipment, installation, and operation procedures.

The controlled reciprocating engines show a reduction in NOx output of approximately 90 percent while the gas turbines yield a reduction of approximately 67 percent.

**TABLE 13-4. NOx Emission Factors for Gas-Fueled Systems.**

| System | NOx (grams/kWh) | |
|---|---|---|
| | Uncontrolled | Controlled |
| Reciprocating IC | 16.2 | 1.4 |
| Gas Turbine | | |
| > 15 MW | 2.46 | 0.82 |
| 4-15 MW | 2.83 | 0.94 |
| < 4 MW | 2.97 | 0.99 |
| Fuel Cell | negligible | |

All data sources generally were in agreement. In fact, for gas turbines the match was almost exact. This appears to be the result of extensive testing performed by the manufacturers in support of the EPA's New-Source Performance Standards Program (NSPS).

## AIR QUALITY IMPACTS

To place the emissions data in perspective, the impact on air quality must be analyzed. Table 13-5 shows a comparison of absolute emissions from a variety of fuel-burning sources. From the data, it is evident that natural-gas-fueled cogenerators produce fewer emissions than many other sources. This comparison is not complete, however, because it reflects source strengths only.

## CONTROL TECHNOLOGY

Equipment manufacturers have developed a variety of control technologies to reduce emissions. Current techniques for reciprocating engines include:

- Modification of the air/fuel ratio;
- Retarding the ignition;
- Air cooling of manifolds;
- Recirculating exhaust gas;
- Derating the engine;
- Modifying the combustion chamber; and
- Utilizing exhaust catalytic converters.

Currently, gas turbines rely exclusively on water or steam injection, although other methods are under investigation.

Because higher combustion temperatures normally increase the formation of NOx, the goal of most NOx-control techniques is to reduce the temperature at which the gas is burned. Manufacturers were surveyed and the literature reviewed to assess the changes in equipment cost, maintenance cost and fuel consumption of the various control techniques. Table 13-6 presents the results.

## TABLE 13-5. Comparison of Emission Rates from Typical Fuel-Burning Sources.

| Typical Source | Fuel | Emissions (lbs/hr) | | | | |
|---|---|---|---|---|---|---|
| | | Particulates | $SO_2$ | $NO_x$ | CO | HC |
| Utility boiler[a] | Coal | 8 | 300 | 150 | 11 | 3 |
| 250 x 10⁶ Btu/hr heat input | Residual[c] | 8 | 200 | 75 | 8 | 2 |
| | Distillate[b] | 4 | 130 | 75 | 9 | 2 |
| | Natural gas | 3 | ◁1 | 50 | 4 | ◁1 |
| Large industrial boiler[a] | Coal | 25 | 300 | 175 | 11 | 3 |
| 250 x 10⁶ Btu/hr heat input | Residual[c] | 25 | 200 | 75 | 8 | 2 |
| | Distillate | 4 | 130 | 75 | 9 | 2 |
| | Natural gas | 3 | ◁1 | 50 | 4 | ◁1 |
| Medium industrial boiler[a] | Coal | 15 | 180 | 114 | 10 | 4 |
| 150 x 10⁶ Btu/hr heat input | Residual[c] | 13 | 150 | 75 | 5 | 1 |
| | Distillate[b] | 2 | 78 | 27 | 5 | 1 |
| | Natural gas | 2 | ◁1 | 27 | 3 | ◁1 |
| Industrial/commercial boiler | Coal | 5 | 60 | 13 | 3 | 1 |
| 50 x 10⁶ Btu/hr | Residual[c] | 4 | 53 | 25 | 2 | ◁1 |
| | Distillate[b] | ◁1 | 26 | 9 | 2 | ◁1 |
| | Natural gas | ◁1 | ◁1 | 9 | 1 | ◁1 |
| Commercial boiler | Coal | 7 | 32 | 3 | 4 | 1 |
| 10 x 10⁶ Btu/hr heat input | Residual[c] | 1 | 11 | 5 | 1 | ◁1 |
| | Distillate[b] | ◁1 | 3 | 2 | 1 | ◁1 |
| | Natural gas | ◁1 | ◁1 | 1 | ◁1 | ◁1 |
| Residential commercial furnace | Distillate[b] | ◁1 | ◁1 | ◁1 | ◁1 | ◁1 |
| ◁1 x 10⁶ Btu/hr | Natural gas | ◁1 | ◁1 | ◁1 | ◁1 | ◁1 |
| Reciprocating IC engine, 1,030 hp uncontrolled (0.725 MW output) | Natural gas | ◁1 | ◁1 | 23 | 23 | 5 |
| Reciprocating IC engine, 1.030 hp controlled (catalytic) (0.725 MW output) | Natural gas | ◁1 | ◁1 | 2 | 11 | 2 |
| Gas turbine (3 MW power output) uncontrolled | Distillate[d] | 2 | 9 | 31 | 2 | 2 |
| | Natural gas | – | – | 21 | 1 | 2 |
| Gas turbine (3 MW power output) controlled (water injection) | Distillate[d] | 1 | 9 | 9 | 7 | 2 |
| | Natural gas | – | – | 7 | 3 | 4 |

[a] NSPS, NSPS of 0.1 assumed for particulate for all industrial (commercial) boilers.
[b] Sulfur content 0.5 percent by weight.
[c] Sulfur content 1.0 percent by weight.
[d] Sulfur content 0.2 percent by weight.

**TABLE 13-6. Effects of NOx Control Techniques for Reciprocating Engines.**

| Technique | Increase in Capital Cost | Increase in Maintenance Cost | Increase in Fuel Consumption |
|---|---|---|---|
| Change Air/ Fuel Ratio | 0 | 3% | 2% |
| Retard Ignition Timing | 0 | slight increase | 3% |
| Air Cool Manifold | 1.5% | 2% | 0 |
| Exhaust Gas Recirculation | 5% | 80% | 0.3 to 0.8% |
| Derating (15%) | 15% | 15% | 0.3 to 14% |
| Combustion Chamber Mods. | $1-47/kW | slight increase | 4% |
| Catalytic Converter | 4%+ | slight increase | 0 |

Water or steam injection is the only technique currently employed on gas turbines. The costs incurred by using this technique were estimated to be three to ten percent of the cost of the turbine, not including any water treatment system that may be required. Fuel consumption at a water/fuel ratio of 1 would increase less than 6.5 percent, but power would also increase 3.6 percent. Operating costs would increase, but the amount would vary based on water cost and treatment requirements.

Although water injection is the method currently available, other techniques being developed include selective catalytic reduction, lean combustion, and staged combustion.

Selective catalytic reduction (SCR) consists of introducing ammonia into the exhaust in the presence of a catalyst. The NOx and ammonia react to form nitrogen and water. SCR is capable of reducing NOx emissions by 80 to 90 percent.

Lean combustion lowers the combustion-chamber temperatures, thereby reducing NOx formation. Several manufacturers have developed experimental combustors with lean-burn configurations. NOx can be reduced by up to 95 percent, although reductions of 50 to

60 percent are more common. The systems currently have control and flashback problems, which prevent commercial introduction.

Staged combustion is a two-step process: combustion begins in a rich zone, where NOx generation is limited by the scarcity of oxygen. Combustion is completed in a lean zone, where the gas temperature is too low for NOx formation. This rich-burn, quick-quench concept is capable of reducing NOx by 70 to 90 percent.

Cost data on these techniques are scarce, but manufacturers contend that they all are more cost effective than water injection. SAI and CH2M HILL, contractors for the Gas Research Institute project which provided the data given in this chapter, are currently performing a detailed life cycle cost analysis of emission control technologies for both turbine and reciprocating engines.

# CHAPTER 14

# Cogeneration in Arizona and California: An Acute Contrast

*R.T. Baltes, W.J. Murphy*

Cogeneration generally can save a facility owner money when the building complex meets the following criteria: 50,000 square feet or more, electrical and thermal energy balance and a portion of the load being of 24-hour duration. A facility with these characteristics is a viable cogeneration candidate today, whether or not the owner sells power back to the local utility company.

In other words, energy cost savings alone are enough to guarantee a payback on a privately owned power plant within a reasonable period of time. The secret to success in most areas of the country today is sizing a cogeneration plant to meet a *portion* of a facility's energy load. The owner generates and buys electricity but he does not sell it.

California is widely recognized as the present-day leader in the field of cogeneration, largely because a statewide shortage of central-plant generating capacity has forced power companies and state regulators to encourage non-traditional energy sources. Our studies show that cogeneration in Arizona is increasingly attractive even though the state has a surplus of energy and state regulators are unenthusiastic about cogeneration as an alternative energy source.

## GEOGRAPHIC DIFFERENCES

A comparison of Arizona Public Service Company (APS) and San Diego Gas & Electric Company (SDG&E) in the neighboring states of Arizona and California demonstrates the extremes in state-imposed incentives and disincentives and their impact on cogeneration.

Three areas account for most of the differences in cogeneration incentives: 1) the fuels used by the local utility, which in turn affect the cost per kilowatt-hour of electricity in the area, 2) the utility's reserve margin and 3) the state regulators' attitudes toward cogeneration.

Figure 14-1, contrasting APS and SDG&E, highlights the incentives and disincentives to cogeneration in Arizona and California. The exhilarating effects of success in California have encouraged a burgeoning cogeneration market there.

| | Arizona Public Service Co. | San Diego Gas & Electric Co. |
|---|---|---|
| No. of QFs | 1 | 48<br>(17 coming online)<br>(47 under discussion) |
| QF Generating Capacity | | 111 MW<br>(52 MW coming on line)<br>(163 MW under discussion) |
| Buyback Rate | 3.6¢ average | 6.5¢ average |
| Backup Rate (Standby) | $49/KW/yr. | $12/KW/yr. |
| Cogen Incentive Gas Rate | None | 50.03¢ per therm |
| Comm/Indus Elec. Rates | 6.4¢ kwh | 11.8¢ kwh |
| Residential Elec. Rate | 7.7¢ kwh | 11.7¢ kwh |
| Utility Generating Capacity | 3,530 MW | 2,300 MW |
| Reserve Margin (1984) | 22% | 32% |
| Fuel Mix | 90.0% coal<br>4.8% gas<br>.5% oil<br>.3% hydro | 30% gas<br>24% oil<br>2% nuclear |
| Purchased Power | 4.4% | 44% |
| Power Sales (Annual) | 12.8 million MWH | 10.4 million MWH |
| No. of Elec. Customers | 469,000 | 834,000 |

FIGURE 14-1. A Comparison of Incentives and Disincentives to Cogeneration

The energy expert must investigate a series of factors when evaluating the feasibility of cogeneration: 1) the costs to interconnect with the utility grid, 2) backup rate, which is the price charged cogenerators for standby power, 3) maintenance rate, the price charged cogenerators to buy power during pre-scheduled downtime for maintenance, 4) supplemental rate, the price of electricity purchased to make up the shortfall between a cogenerator's capacity and energy load, and 5) buyback rate.

Buyback rate is also called "avoided cost" because it is based on the amount of money the utility saves (avoids spending) when the cogenerator provides energy.

The three-member Arizona Corporation Commission has established guidelines regarding buyback rates but as yet there has been little or no experience with cogeneration.

The five-member California Public Utility Commission is widely reputed for its strong activist role in supporting cogeneration. Cogenerators know going in what the rules are (regarding buyback rates and other incentives) and have a much greater chance of developing cost-effective plans for cogenerating all or part of their power needs.

## HELP FROM PURPA

In 1978 Congress enacted PURPA, short for Public Utility Regulatory Policy Act, to assist would-be congenerators. The Federal Energy Regulatory Commission (FERC) drafted guidelines for implementing PURPA. Based on these guidelines, rules were established by state utility regulatory agencies for would-be cogenerators and utilities. The rules vary vastly from state to state, reflecting independent interpretations of the guidelines.

PURPA says cogenerators do not have to supply all of their own electricity, and in addition, requires utilities to buy electricity back from cogenerators. Prior to PURPA, a cogenerator needed a redundant onsite system to ensure an uninterrupted power supply. The biggest advantage of PURPA is that a cogenerator does not have to match exactly his generation to his load. The match can be an economic one with the supplemental power provided by the utility. The key is to size the cogeneration system to meet the minimum thermal load. In most states the resulting electrical capacity should be less than the minimum electrical load.

Utility buyback at attractive rates allows a cogenerator to size his onsite plant to meet the facility's peak energy demands and to sell any excess generating capacity back to the utility company. Buyback rates (particularly high priced ones) are unstable and do not have the predictability of retail rates.

Figure 14-2 shows a typical electrical load profile and three options for sizing a private cogeneration plant. To be totally self-sufficient a plant must match a facility's peak or "maximum" load. Excess electricity generated during the off-peak is sold back to the utility company. Meeting maximum demand is not a viable option in either California or Arizona. California plants generally are sized to meet "optimum" load, thus replacing a significant portion of the utility electricity. Excess electricity is sold back to the power company. Because buyback rates in Arizona are still very low, private cogeneration plants must be sized to meet "base" load requirements with all power generated being consumed onsite.

According to a recent survey by Claire Wooster and Eric Thompson,* PURPA is working best in states where the rules are clear. It is not surprising that most qualifying facilities are located in states where a large degree of certainty exists regarding regulatory climate.

A qualifying facility (QF) is one certified by FERC as having met the requirements of PURPA and thereby qualified for all the law's benefits.

Wooster and Thompson reviewed the orders and hearings of 22 states that chose, either through general rulemaking or the hearing process, to direct in some detail the procedures used by utilities in setting buyback rates. They discovered that two-thirds of the QFs filing with FERC in the past 3 years came from California, Florida, Tennessee and Texas, and that two-thirds of the total capacity offered by new cogenerators is produced in California, Texas and Massachusetts.

The California PUC not only has set favorable buyback rates but also permits cogenerators the same rate for natural gas as electric utilities pay. In addition, it gives cogenerators priority over other industrial users in the case of natural gas shortages. The natural gas incentives appear to be particularly significant because more than three-quarters of California's QF filings list natural gas as their fuel.

*See Chapter 11.

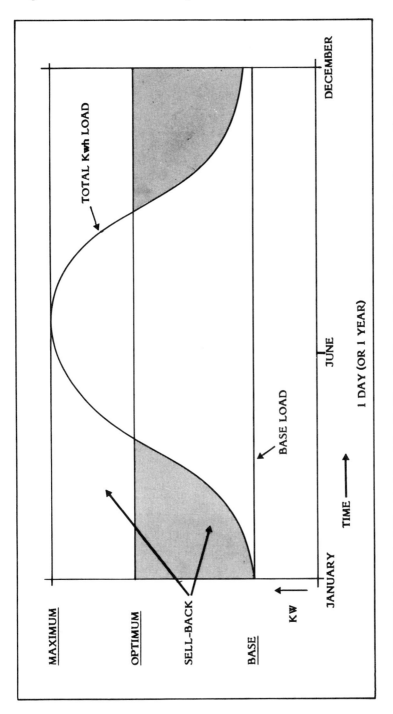

FIGURE 14-2. Typical Electrical Load Profile Showing Three Options for Sizing Private Cogeneration Plants, Depending on the Attractiveness of Utility Buyback Rates and Other Incentives.

APS currently has only one QF compared with at least 65 QFs that will be online in SDG&E's territory by the end of 1984.

What accounts for this large difference? We have touched on the fact that California, especially southern California, is short on low-cost electricity while Arizona's generating capacity is quite healthy. One result is that the retail electric rate in SDG&E's area is 11.8¢, almost double APS' 6.4¢.

## COGENERATION INCENTIVES

One rule of thumb, which is generally reliable, says that when the first digit of the cost of fuel (expressed in dollars per million BTU's) is about half of the cost of the utility-furnished electricity (expressed in cents per kilowatt-hour), cogeneration will probably make economic sense. In the case of SDG&E, the cost of gas is $5 per million BTU and the cost of furnished electricity is approximately 11.8¢ per kilowatt-hour. This fits the rule of thumb and the return on investment is between 20-30%. At APS the cost of gas again is approximately $5 per million BTU, but the cost of furnished electricity is 6.4¢ per kilowatt-hour, making the return on investment marginal.

California utility regulators discourage nuclear and coal plants and are promoters of alternative energy technology. Because the state has a high growth rate, regulators must deal with an increasing population's energy demands and cogeneration has become an available means of producing substantial quantities of electricity—more than 160 megawatts of cogenerated electricity in SDG&E's territory by year's end.

In addition, SDG&E's buyback rate is 70% higher than the APS buyback rate and the California utility's backup rate for standby electricity is only one-fourth the cost of standby electricity from APS.

Other incentives include power companies' willingness to pay half the cost of cogeneration feasibility studies up to a maximum of $25,000. For example, Southern California Edison Company recommends hiring an engineering consultant to perform a feasibility study of electricity-related modifications that show promise of energy and monetary savings. The study concentrates on experimental technologies or applying a proven technology to a new situation.

On the other hand, some power companies exert every effort to discourage the cogenerator. They may charge from $10,000 to

$25,000 to have their planning departments determine how much it will cost to interconnect with their distribution network.

In the case of APS, which owns the natural gas distribution rights in the same area as the electric distribution rights (as does SDG&E), the utility may charge to bring a gas line to the property and yet not sponsor any cogeneration studies.


## THE KEY IS BACKUP RATES

Although reserve margins are important in arguing the need for higher capacity charges in the buyback rate, the most important item in Arizona and recently in California is not the price the utility avoids (buyback rate) but the price the commercial and industrial customer avoids—the retail price of electricity. The price in Arizona (6.4¢) is expected to increase significantly because of the Palo Verde Nuclear Generating Station. Prices in Arizona are expected to increase well over 50% in the next 2 years. This will result in retail prices of at least 10¢ per kWh.

The future for cogeneration then is in replacing utility power (it is easier to predict its price and little chance it will decrease) rather than trying to sell to the utility.

The problem in many areas is not selling electricity back to the utility but instead being able to buy standby electricity from the utility at nondiscriminatory rates for those times when the onsite plant is down for service.

Often the feasibility of cogeneration is judged solely by the buy-back rate that the utility company is willing to pay the cogenerator for excess electricity. Rather than buyback rate, backup charge for standby power appears increasingly to be the key factor.

As the cost of electricity generated by large coal and nuclear plants continues to rise, the cost of energy generated by onsite co-generation plants becomes more competitive, indeed less expensive than utility electricity. By replacing power purchased from the utility with onsite-generated power a savings can be realized. Because of this new reality, cogeneration in Arizona and states like Arizona may well be a valid energy alternative providing standby electricity which can be purchased at reasonable rates.

## ACE IN THE HOLE

A strong probability exists that the climate for cogenerators in Arizona will improve in the near future. Two additional members are to be added to the corporation commission, improving the likelihood of support for renewable energy sources and cogeneration. And there is growing uncertainty about the future of Palo Verde nuclear plant. The high cost per kilowatt-hour for electricity, if the nuclear plant does come online, may force user exodus. Add to this energy-users' increasing awareness of cogeneration's success in California and other states, and popular demand cannot be far behind.

One reason buyback rates are so low in Arizona is that utilities seldom run expensive peaking units because available capacity is so great. But that is expected to change too, and when it does, the door will be open for cogenerators to rush in. Until such time, co-generators' "ace in the hole" is simply to replace the purchase of utility power with less expensive onsite cogenerated energy. Private cogeneration plants could become even more attractive when they begin selling power back to the utility thereby shortening further the payback period on their capital costs.

## INTERIM GAME PLAN

For the present, potential cogenerators in Arizona and states like Arizona should follow a game plan somthing like this: Hire a competent local consultant to evaluate the feasibility of cogenerating, whether or not power is sold back to the utility. But don't negotiate too long. Plan to take your case to the regulatory commission within a designated period of time. Both parties—utility and cogenerator—will be provided the opportunity to present their cases to the commission for a final decision.

Be ready to add cogeneration when the rates for purchased electricity begin to skyrocket, as Arizona Public Service and many other utilities are predicting they will. Realize it takes 18 to 24 months to bring a large cogeneration installation on line, and 6 to 12 months to bring a plant less than 100 kilowatts on line.

In the final analysis, a strong case can be made that progress in California and the rapid acceptance of cogeneration projects in that state have formed the single most positive influence in shaping new attitudes toward the relationship between utilities and industry.

# CHAPTER 15
# A Solar Cogeneration System

*F.J. Krause, E.J. Ney*

The solar Total Energy Project (STEP) at Shenandoah, Georgia, is a cooperative effort between the United States Department of Energy (DOE) and the Georgia Power Company to further the search for new sources of energy.

A part of the National Solar Thermal Energy Program, initially funded by DOE, the Shenandoah Project, shown in Figure 15-1, is the world's largest industrial application of the solar total energy concept. The objective of the Project is to evaluate a solar total energy system that provides electrical power, process steam and air conditioning for a knitwear factory (operated by Bleyle of America, Inc.). Solar energy will generate a large part of the electricity and displace part of the fossil fuels normally used to run the factory and produce the clothing.

Construction of the system was completed early in 1982, when operations were initiated. Solution of unexpected electrical and mechanical problems produced significant information for subsequent system designs. This chapter presents an overview of the Project; a brief System Description; a discussion of various anomalies, together with subsequent high quality solar and thermodynamic system performance results; and the formal, specific tests to be run and thoroughly analyzed in the Test Operations Phase.

## HISTORY

In 1977, DOE declared Georgia Power Company the winner among 16 competitors from 14 states for the location and application of the Solar Total Energy Project. The Georgia Power site most nearly met all project requirements regarding weather, accessibility, energy requirements and other important considerations.

**FIGURE 15-1.** Aerial View of Completed STEP Site.

Design work for the solar energy system was completed between 1978 and 1980. Georgia Power provided cost-sharing support and coordination throughout the design and construction stages, and assumed responsibility for operation of STEP in July, 1982. It is anticipated that ownership will be transferred to the Georgia Power Company near the end of DOE operational funding in July, 1984. The complete STEP schedule is shown in Figure 15-2.

In the future, as full owner of the Project, Georgia Power will operate the facility as part of a Solar Center that will include further land acquisitions, additional offices, visitor space, exhibits, and test facilities for research and development of solar energy technology.

The 8-year period comprising site dedication, design, construction, operation, and dissemination of cost and performance data (shown in Figure 15-2) will culminate with a period of commercial operation that will complete the original 10-year cooperative agreement signed with Georgia Power in May, 1977.

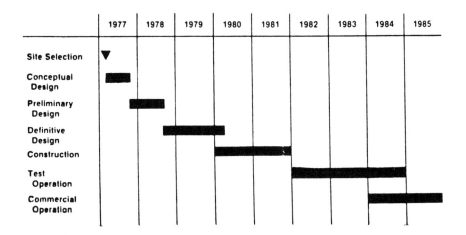

**FIGURE 15-2. STEP Program Schedule Toward Commercial Operation**

## TOTAL ENERGY CONCEPT

The total energy concept—also called cogeneration—makes maximum use of waste heat from electrical power generation to meet other energy requirements. Combined with a solar energy system, the total energy concept offers these benefits:

1. It provides energy from a renewable source.

2. It makes maximum use of collected energy.

3. Its closed-loop system releases no pollution.

4. It is compatible with existing utility services.

## COMMERCIAL APPLICATION

The 25,000-square-foot Bleyle Knitwear Factory has been operating since 1978 with conventional energy sources. When the solar system is fully operational, it will be capable of generating 11 billion BTU annually (11 million BTU per hour peak thermal energy), which can be translated into 400 kW of electricity, 1380 pounds per hour of process steam at 350 degrees F and 120 psig, and 257 tons of air conditioning, which can be supplied to the knitwear factory. Energy needs beyond the solar derived portion required by the Bleyle Plant will be supplied by conventional sources.

The Bleyle Plant building was designed to include DOE and Georgia Power recommendations for achieving energy efficiency:

1. Reduced height to minimize volume of building and wall area.
2. Four-foot insulating earth bern as thermal buffer around the building.
3. North-south orientation.
4. Air conditioning economizer cycle.
5. Super insulated walls and roof.
6. High efficiency fluorescent lighting.
7. Energy efficient equipment.

Energy-conserving features alone, exclusive of the solar equipment, have reduced the factory's energy needs by 46%, thus saving more than $25,000 a year (at 1983 utility rates). Data gathered by Georgia Power instruments in the factory were used to determine the building's energy requirements, and this information was used to design the solar energy system.


## SITE DESCRIPTION

The aerial photograph in Figure 15-1 shows the physical layout of the project site. A field of 114 parabolic solar dish collectors—each 23 feet in diameter—tracks the sun and concentrates the rays to heat a circulating fluid. An easement obtained from adjacent landowners guarantees unobstructed sunlight for collectors.

The Bleyle Knitwear Plant is shown in the upper left corner of the photo. The white building at the upper left corner of the collector field houses the operations and control equipment, and the area to the right of the building contains the steam generator, high temperature storage tank, and other mechanical equipment.

A meteorological station at the Site, operated by the Georgia Institute of Technology, constantly monitors the amount of solar energy available. The solar insolation and surface weather instruments make it one of the most sophisticated stations in America for gathering data about the sun. Information collected by the station was used in designing the Solar Total Energy System and will continue to be used to support the national weather network.

A Georgia Power electrical substation (upper right) designed for the Shenandoah project is providing new technology and engineering experience for integrating the electrical output of the cogeneration solar system with the company's 15,000-plus megawatt system.

## PARTICIPANTS

The organization show in Figure 15-3 participated in the development of the Shenandoah Solar Total Energy Project over the first 5 years of planning, designing, construction, initial operation, and testing. The diagram illustrates the relationships among the participants.

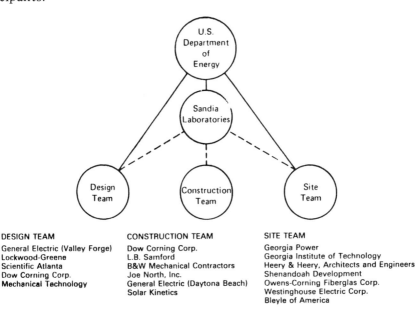

DESIGN TEAM
General Electric (Valley Forge)
Lockwood-Greene
Scientific Atlanta
Dow Corning Corp.
Mechanical Technology

CONSTRUCTION TEAM
Dow Corning Corp.
L.B. Samford
B&W Mechanical Contractors
Joe North, Inc.
General Electric (Daytona Beach)
Solar Kinetics

SITE TEAM
Georgia Power
Georgia Institute of Technology
Heery & Heery, Architects and Engineers
Shenandoah Development
Owens-Corning Fiberglas Corp.
Westinghouse Electric Corp.
Bleyle of America

**FIGURE 15-3. Overall Organizational Relationship Among Participants in Solar Total Energy Project.**

## PROGRAM OBJECTIVES

The overall U.S. Department of Energy (DOE) objectives for the National Solar Total Energy Project at Shenandoah, Georgia, are to:

1. Produce engineering and development experience on large scale

solar total energy systems as preparation for subsequent commercial size applications.

2. Assess the interaction of solar energy technology with the application environment.

3. Narrow the prediction uncertainty of the cost and performance of the Solar Total Energy System (STES).

4. Expand solar engineering capability and experience with large-scale hardware systems.

5. Disseminate information and results.

## SITE OBJECTIVES

The primary objective of the STEP Site/Application effort at Shenandoah is to provide a commercial facility to utilize solar derived electrical and thermal energy, as well as a suitable area for erecting a solar energy system to provide the required energy to the facility. This includes data acquisition and analysis, as well as design interface.

The objectives of the Georgia Power Company within these parameters are to:

1. Evaluate the significance of an emerging alternate energy technology in an industrial application.

2. Promote the utilization of energy conservation and load management.

3. Consider the applicability of cogeneration facilities.

4. Analyze the economic potential of solar total energy in an industrial application.

The achievement of these objectives will allow Georgia Power Company to better provide reliable, economic and environmentally acceptable energy to the consumers of the State of Georgia and help lead the nation to a partial solution of the energy dilemma.

## PROJECT DESCRIPTION AND SYSTEM DESIGN

The Solar Total Energy Project is in Shenandoah, Georgia, 25 miles southwest of Atlanta International Airport, at Exit 9 of Interstate 85, as shown in Figure 15-4. The Site consists of 5.72 acres in the Shenandoah Industrial Park adjacent to and east of the Bleyle Knitwear Plant.

**FIGURE 15-4. Georgraphic Location of STES Site.**

## Solar Energy System Description

A solar total energy system uses collected solar energy to supply high-grade electrical and mechanical energy and low-grade thermal energy for selected applications. The Solar Total Energy Project (STEP) at Shenandoah supplies electric power to a utility grid, and process steam and air conditioning to a knitwear manufacturing facility. Excess power from the STEP is supplied to the Georgia Power Company electricity distribution network.

The STEP is a fully cascaded total energy system with parabolic dish solar collectors and steam Rankine cycle power conversion system capable of supplying 100-400 kWe output with process steam extraction. The design includes the Solar Collection Subsystem, the Power Conversion Subsystem, the Thermal Utilization Subsystem, and the Control and Instrumentation Subsystem, which are monitored to provide the data necessary to evaluate the STEP. Figure

15-5, a simplified schematic diagram of the STES, illustrates the overall concept of cogeneration with solar energy.

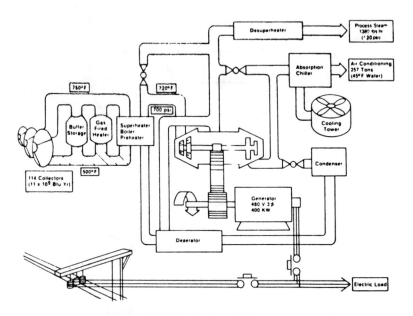

**FIGURE 15-5. Illustration of STES Concept.**

## Operation

Operation of the system begins with circulation of a heat transfer fluid through the receiver tubes of the parabolic dish solar collectors. Solar radiation is focused in the receivers by the collector reflector and heats the silicone heat transfer fluid (HTF) to 750°F. The heat transfer fluid is then pumped to a heat exchanger.

In the heat exchanger, the heat transfer fluid boils water and superheats the steam; the heat transfer fluid then returns to the collectors and the cycle is repeated. The superheated steam drives a turbine that in turn drives an alternator. Steam at 350°F is extracted from the turbine for knitwear pressing. The low-pressure steam exhausted from the turbine is used to produce chilled water for air conditioning, or is cooled as it passes through an air-cooled condenser.

## Solar Collection Subsystem

The Solar Collection Subsystem (SCS) consists of an array of 114 parabolic dish collectors, each 23 feet in diameter, shown in Figure 15-1. The heat transfer fluid flowing through the collectors, whose receivers are connected in parallel, is heated from the inlet temperature at 500 degrees F to 750 degrees F. The receiver is a cavity type capable of receiving an incident concentrated solar flux equal to 235 suns. The concentrated solar flux impinges upon the receiver coil's absorptive surfaces enclosed within the insulated cylindrical shell.

Each parabolic dish is made up of die-stamped aluminum petals and was assembled in the field. The aluminum petal is laminated with a second surface—aluminized acrylic reflective film—prior to forming. Each collector tracks individually in polar and declination axes to follow the sun from morning to evening, and from season to season. The parabolic dish collectors are arrayed on the Shenandoah collector field in a repeating diamond pattern.

The field piping network consists of welded pipes in the main manifolds, and steel tubing in the branches. All are covered with a high-temperature insulation. The SCS provides one hour of thermal storage at 750°F as a buffer against transient solar conditions. Energy is stored in the silicone heat transfer fluid in a thermocline tank. A natural gas fired heater capable of supplying the power conversion subsystem energy input requirements is used during startup and to supplement the solar energy system as necessary.

## Power Conversion Subsystem

The Power Conversion Subsystem (PCS) consists of a three-piece pool-type boiler with preheater, boiler and superheater, a steam turbine alternator rated at 500 KVA, an air-cooled condenser and condensate storage tank, make-up demineralizer, deaerator, and necessary pumps. In normal operation, steam at 720°F and 700 psig is generated in the boiler-superheater and delivered to the turbine inlet.

The turbine alternator consists of a four-stage, high-speed (42,450 rpm) turbine, a gearbox that reduces the speed to 1800 rpm, and a 60 Hz alternator. The low pressure side of the high pressure turbine stages has an extraction port for process steam and steam for regenerative feed water heating. The low pressure turbine stages exhaust into an air-cooled condenser at 230°F and provides steam to the Thermal Utilization Subsystem (TUS).

### Thermal Utilization Subsystem

The Thermal Utilization Subsystem serves as the condensing medium for steam and the energy source for cooling the Bleyle knitwear factory. Steam enters the subsystem from either the turbine low pressure exhaust or when the turbine is out of service, through a pressure reducing turbine by-pass valve.

Steam from either the turbine or the by-pass valve is routed to an absorption chiller and an air-cooled condenser. Chilled water produced by the absorption machine provides cooling for the Bleyle factory. After the steam requirements for the absorption machine are satisfied, any excess steam is circulated through the air-cooled condenser. Condensed water from the absorption machine and the condenser are then returned to the hot well or a condensate storage tank where they may be recycled.

### Control and Instrumentation Subsystem

The STEP control system provides a full range of operations covering minimum operator control to extensive data collection for analysis of experimental operations. The control system partitions control functions between a mini-computer and its peripheral equipment and micro-processors distributed through the system. These micro-processors exercise some control functions locally.

The Collector Control Units (CCU) located at each collector and the Control and Instrumentation Subsystem (CAIS) are connected by redundant serial links. This allows communication among the distributed control system components by a single pair of leads. Other sensors, including the weather instruments, interact with the central control computer through the Energy Utilization Processor (EUP).

The CAIS provides the following:

1. Control of all subsystems and components for normal and fail-safe operations.

2. All control logic for operational modes as selected by the operator.

3. Collection, monitoring of data and processing of information for: Automatic decisions; System status information; Stored information for subsequent system analysis and evaluation; and, Automatic initiation of safety measures to prevent hazards to personnel, damage to equipment and property, and loss of data.

In addition, operator control is provided for experimental modes to characterize system and component performance over ranges of operational parameters and to identify operating strategies for more effective electric and thermal energy displacement. The switch to the experimental modes allows the operator to initiate solar collection experiments, and to monitor the record data. Diagnostic routines may be initiated in the event of a malfunction.

The CCUs perform the following functions:

1. Receive system control information from the CAIS and provide signals to collector field control equipment, such as drive motors and valves.
2. Interpret local data to identify potential hazards and initiate control actions to avoid or minimize damage to the collector.
3. Maintain proper sun tracking automatically once adequate focus by central computer has been established.
4. Relay data from local instruments to the CAIS for further processing or storage.

## System Loads

The STEP loads include electric loads and process steam and cooling for the knitwear manufacturing facility. The design loads used to size the STEP are summarized in the following table. Except for lunch and shift breaks, the knitwear manufacturing facility electrical load profile is relatively constant over a one-shift operation. Process steam at saturated conditions is required during all working hours.

**Peak Load Requirements for Knitwear**
**Manufacturing Facility**

|                                    |                            | *STEP Capacity*            |
| ---------------------------------- | -------------------------- | -------------------------- |
| Electrical                         | 161 kW                     | 400 kW                     |
| Cooling                            | 1420 Mj<br>(113 tons)      | 3260 Mj<br>(257 tons)      |
| Process Steam<br>(117° C, 350°F)   | 626 Kg/hr<br>(1380 lbs/hr) | 626 Kg/hr<br>(1380 lbs/hr) |

The cooling loads consist primarily of internal heat generated by the process steam machinery, people, and building lighting and are relatively constant during plant operating hours. The plant's heating, ventilating and air conditioning (HVAC) system incorporates an economizer cycle that supplies a portion of the cooling load during the winter months. The cooling loads are met by a chilled water system supplied by an absorption chiller.

The total number of heating degree days for Atlanta is 3,095, and the total number of cooling degree days is 1,595 (using 65 degrees F as a base). The heating season generally extends from October to April, with occasional heating required in May and September. The cooling season extends from May through September with occasional cooling required during March, April and October. This balanced situation allows research data gathered at Shenandoah to be generalized for much of the United States.

## SYSTEM PERFORMANCE

### Anomalies and Achievements

Construction of the solar project was completed early in 1982. During that year, startup operations were conducted by a joint operational team of Sandia National Laboratories and Georgia Power Company. Various unexpected electrical and mechanical problems provided significant information for subsequent system design applications. These problems have now been resolved, and the program has moved into the experimental operations test phase. Following is a summary of major events of 1982.

The STEP Steam System Integrity Tests were completed on January 21 with the first synchronization and generation of electricity to the 100 kW level. Manual Control of the Balance of Plant (BOP) was also achieved during the month.

In February, extracted steam was supplied from the turbine to the Bleyle Plant for dryout of thermal insulation that had been dampened during construction activities. This operation was carried out under manual control of the BOP.

In March, the major activity was a two-day inspection of the STEP by a formal Readiness Review Committee. No major problems were identified by the committee, but some recommendations were provided. The original motors and potentiometers for each solar collector were removed and waterproofed due to their failure rate in a high-rainfall area. By the end of March, construction was essentially completed at the site.

In April, the project's operational status was reviewed by the Department of Energy, Sandia, and Georgia Power. The group gave provisional acceptance based on resolution of specific problems. The major anomaly was operation of the Control and Instrumentation Subsystem (CAIS), with the Collector Field Subsystem (CFS), and the Balance of Plant (BOP). This computer-related problem was resolved late in the year, when a major design change was made.

A major milestone was achieved on May 10, when the site dedication was held. More than 500 people attended the formal ceremonies, with Georgia Power Chief Executive Officer Robert W. Scherer, DOE's Dr. Robert San Martin, Congressman Newt Gingrich, Sandia's Don Schueler, and Atlanta Mayor Andrew Young as participants.

On May 12th, the turbine-generator was synchronized to a load of 200 kW and solar-generated electricity was produced for the first time. After many problems over a 5-month period with the condensate pump, the manufacturer inspected the unit on site and found it to be running backward due to a directional flow arrow on the casing that had not been clarified. The pump was then made fully operational.

In June, the air conditioning capability of the project was demonstrated. After the integrity of the chilled water system to the Bleyle Plant was verified, the absorption chiller was started. Several days of successful operation of this air conditioning system led to a major milestone on June 15th: cogeneration with approximately 250 kW (electric) and 50 tons air conditioning (thermal).

A leaky accumulator was removed from the boiler feedwater pump discharge. Inspection proved that the Viton bladder would not accomodate a working temperature of 330°F. To correct this problem, an accumulator of a piston design with Vitron O-rings was purchased but this problem resulted in 17 days of downtime.

In July, the reflective film on 15 solar collectors was damaged. Stray concentrated light from adjacent collectors had been focused on the backs of the damaged collectors. The repair of the film was completed within the month.

Process steam was provided to the Bleyle Plant for processing needs for the first time in August. Process steam (5600 pounds) and air conditioning (600 ton-hours) were provided by semi-manual operation of the solar collectors. On August 20th, by pumping heat transfer fluid considerably cooler than the steam generator temperature, the steam generator was thermally shocked causing leaks at the tube-tubesheet interface. However, this problem was not completely identified and corrected until December.

In September, a new type of seal was used to replace the tungsten carbide shaft seal on the steam generator HTF pump. This original type of seal had failed seven times since the pump had been installed. The new silicon carbide seal operated without problems.

Also during September, it was determined that the Central Processing Unit (CPU) memory of the 128K-word computer would be inadequate to handle multiple subsystems by CAIS. It was decided to change to a 256K-word unit to handle the operation of all the subsystems efficiently. On September 23rd, for 7.5 hours, $10 \times 10^6$ Btu of solar steam was produced through computer control of the collector field by the CAIS.

During October, the Balance of Plant (BOP) was operated for a significant time in the thermal energy mode. On the 15th, a typically good solar autumn day in Georgia, the solar collectors auto-tracked for 9 hours and 23 minutes, providing more than $30 \times 10^6$ Btu of energy. Approximately 72% of the energy delivered to the Bleyle Plant that day was solar derived. A total of 61040 lbs. of process steam and 4048 ton-hours of air conditioning were supplied to the knitwear plant for the month.

The new computer was received in November. The STEP staff, with assistance from General Electric and Auburn University, preceded the changeover with debugging efforts, checkout of the operation programs, and the creation of data analysis programs. In addition to the computer work, repair of the steam generator was initiated. The problem became apparent when water was discovered in the heat transfer fluid.

The steam generator leak was repaired in December. A redesign of the nitrogen supply system, using a 315-gallon bulk liquid tank, resulted in an operating cost savings of $10,000 per year. A larger capacity air compressor was selected due to the inadequacy of the original, which became a spare. Computer control of the plant, using the new computer, was achieved in late December.

These electro-mechanical problems provided valuable experience to the design and operation data base necessary for subsequent designs. Each subsystem has performed at its design level, and the total performance of STEP is now being evaluated. Highlights of this performance follow.

### Performance Highlights

With the initiation of test operations, significant solar data were collected and reported. The following table shows the experimental data for direct solar radiation, and compares the data to the solar model data produced by the Georgia Institute of Technology.

**Comparison of Direct Solar Radiation**

| 1983 | Experimental Data Btu/Sq Ft/Day | Solar Model Btu/Sq Ft/Day |
|---|---|---|
| Jan | 947 | 926 |
| Feb | 839 | 1328 |
| Mar | 1059 | 1444 |
| Apr | 1492 | 1602 |
| May | 1098 | 1675 |
| June | 991 | 1547 |
| July | 1046 | 1602 |
| Aug | 700 | 1543 |
| Sep | 544 | 1471 |
| Oct | 1314 | 1587 |
| Nov | 880 | 1511 |
| Dec | 324 | 1163 |
| **Total** | **11234** | **17399** |

The integrated amount of energy measured in 1983 is 35 percent less than the solar model predicted. This significant decrease in available sunshine has resulted in lower than expected usable energy values for the output of the STEP. The reasons for the low measured values may be associated with expected statistical variations or reductions due to the Mexican Volcano, El Chichon, or could be errors in the method used for obtaining the direct insolation from the solar model. These possibilities are being investigated.

Overall, the Shenandoah Solar Total Energy Project has achieved the following major objectives:

A significant number of engineers, scientists, and students have been trained and validated as STEP operators and data analyzers.

All major thermodynamic components of the project have been proven and have met their design values. These include the steam turbine generator, the collectors, the absorption chiller, and the high temperature storage system.

Solution of the anomalies in the small mechanical components—such as motors, pumps, potentiometers, and valves—has provided significant information for future advanced system designs.

Additional checkout time and effort applied to the hardware and software aspects of the control and instrumentation system has provided a significant base for more efficient designs and checkout of future systems.

Formal testing for a variety of test modes has been initiated and will take approximately one year to complete and report.

The U.S. Department of Energy and Bleyle Corporation of America, STEP's two primary "customers," have documented their satisfaction in working with Georgia Power and have reported that all their objectives and requirements have been met safety and efficiently. The STEP is operated under contract to DOE, which receives the experimental data, and Bleyle receives the energy output from the project.

On March 8, 1983, a steady state operation was achieved for one hour. During that time, a solar collector field efficiency of 52.6 percent was measured. The calculated solar collector efficiency is imposed on the following waterfall chart. The field thermal operating losses are calculated to be 9.3 percent of the energy at the solar collector receivers, which is considerably better than was calculated for the original design.

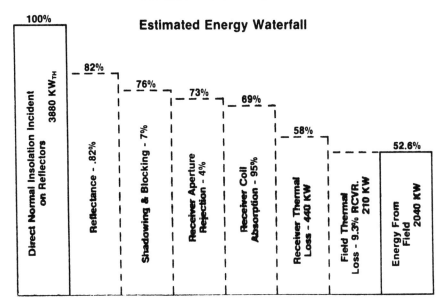

Collector Field Subsystem

**Estimated Energy Waterfall**

On May 17, 1984, steady state operation for one hour was achieved during the formal performance testing program. The test required that fossil energy be used for the source. Energy outputs were electricity and process steam. The calculated efficiencies are imposed on the waterfall chart below.

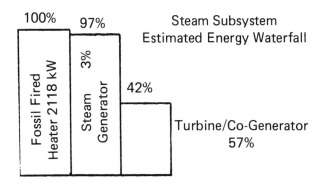

Steam Subsystem
Estimated Energy Waterfall

During the checkout phases of the project, conducted primarily through 1982, and for the beginning of the performance testing, conducted in 1983, useful production data were collected. Usable energy produced from the beginning of operations through 1983 is shown below for the three forms of energy produced by STEP and used by the Bleyle Knitwear Factory:

**STEP Production Summary**
**(through December, 1983)**

| | |
|---|---|
| Electricity (kWhr) | − 234287 |
| Air Conditioning (ton-hrs) | − 62515 |
| Process Steam (pounds) | − 892399 |

## CONCLUSION

The Solar Total Energy Project, sponsored by DOE and Georgia Power, provides a major stepping stone for industrial solar energy development. Since January of 1982, when plant startup and acceptance tests began, a great deal of valuable information for succeeding designs has been obtained through solution of various electricity and mechanical problems.

Collection of plant performance data early in 1983 has demonstrated that the primary systems perform thermodynamically as projected during the design phases. The plant is continuing formal test operations by the Georgia Power operating team. A minimum of 29 tests will be performed out of a total of 78 that were developed for us by EPRI with input from Sandia/DOE and Georgia Power. The resulting analysis of these tests will be the basis for the planned Commercial Operating Phase scheduled for 1985 and 1986.

Also, the results of a STEP Reliability Study by Auburn University will provide a prediction of system availability for various operating conditions. When the full schedule is completed, system owning and operating costs, system and subsystem performance, and lessons learned should provide the bottom line answer: when and under what conditions will this generic solar system be commercially cost effective.

# CHAPTER 16

# Site Selection, Design, and Data Analysis of An Onsite Fuel Cell Cogeneration System

*R.W. Taylor, Lt. J.P. Fellner*

The installation of a 40-kW fuel cell power plant at the Air Force Museum, Wright-Patterson Air Force Base, Ohio, is the first of three installations at military bases scheduled for 1984. The U.S. Department of Defense (DOD) and the Department of Energy (DOE) are jointly funding these installations. In addition to Wright-Patterson AFB, fuel cell power plants will be installed at Sheppard AFB in Wichita Falls, Texas, and at Ft. Belvoir near Alexandria, Virginia.

The NASA Lewis Research Center (LeRC) was selected to administer the DOD field test and Science Applications, Inc. (SAI) was selected as the contractor responsible for installation of the fuel cell power plants and associated data acquisition equipment.

The Air Force, through the Air Force Wright Aeronautical Laboratories, is developing phosphoric acid fuel cell power plants for remote site applications where utility power is not available. These applications consist mainly of communications and radar sites and typically use diesel or gasoline generators as their power source. They are located around the world in hard-to-reach sites, with climates ranging from arctic cold to desert heat. This results in a high delivered fuel cost for the generators.

Since the fuel cost is high and many of the sites are hard to reach, efficient and reliable electric generators can greatly reduce the life-cycle-costs for the power systems. After studying many different technologies, fuel cell power plants were selected as the best power system candidate to meet these requirements. The purpose of the field test at the Wright-Patterson AFB is to evaluate the 40-kW fuel cell power plant and determine its ability to meet military remote, mobile, and facility requirements.

The installation of 40-kW fuel cells at military bases is a separate, but parallel, program to the Onsite Fuel Cell Field Test Project. The latter project is a multi-year effort which is jointly funded by the Gas Research Institute (GRI), DOE, and over 30 gas and electric utilities throughout the United States including two gas utilities from Japan. GRI is managing the project, United Technologies Corporation (UTC) is the fuel cell power plant manufacturer, and SAI is serving as the overall coordinating contractor. The NASA LeRC was selected by the DOE to manage the manufacturing contract with UTC.

The GRI Field Test Project is composed of field testing of over forty 40-kW fuel cell power plants and a market/business assessment of commercial onsite fuel cell power plants. Each utility will field test at least one power plant and will conduct a market/business assessment of commercial fuel cell service.

## SITE SELECTION

Initially a study was conducted to evaluate the energy conservation and economic potential of onsite fuel cell energy service for the Air Force. Thirteen facilities at three Air Force Logistic Command Bases were identified as possible test sites for the onsite fuel cell energy system. The selection was based on the compatibility of the thermal energy usage at the site with the thermal energy produced by the fuel cell and an estimate of the projected operating savings of electric and fuel costs for each site. The following site selection criteria were used:

- The site should be representative of a class or type of application common to the Air Force.

- The site should be suitable for display and demonstration to selected Air Force personnel and the public. Consideration should be given to aesthetics, safety and visitor facilities.

- The fuel cell site should be reasonably close to the site thermal load. The thermal load should be temperature compatible and as nearly continuous as possible (approximately 150,000 Btu per hour or nearly 110 million Btu per month). Thermal storage may be used to improve coincidence.

- The application should have a critical load requirement of less than 40 kilowatts for backup emergency power.
- The voltage and other electrical interface requirements should be compatible with the power plant specification.
- A natural gas supply should be near the site.

Based on these criteria, nine candidate sites were selected at the Wright-Patterson AFB for possible installation of a 40-kW fuel cell power plant. The following is a list of these facilities and corresponding thermal applications for the heat produced by the fuel cell:

| *Facility* | *Fuel Cell Thermal Use* |
|---|---|
| Headquarters | Domestic Hot Water |
| Officers Quarters (855) | Domestic Hot Water |
| Officers Quarters (856) | Domestic Hot Water |
| Gymnasium | Domestic Hot Water and Pool Heating |
| Museum | Space Heating |
| Heating Plant | Feedwater Heating |
| Cafeteria | Domestic Hot Water (incl. Dishwashing) |
| Dining Hall | Domestic Hot Water (incl. Dishwashing) |
| Medical Center | Domestic Hot Water |

Several sites at other Air Force bases were also evaluated. The Air Force Museum was Wright-Patterson's preferred choice for installation of a 40-kW fuel cell power plant.

## FIELD TEST

A 40-kW fuel cell power plant was installed at the Wright-Patterson AFB in August 1984. The associated Data Acquisition System (DAS) also installed at this time is discussed later in this chapter. SAI was responsible for the installation and start-up of the power plant. With assistance from SAI, Wright-Patterson AFB personnel will operate and perform normal maintenance on the power plant.

The fuel cell is expected to operate for one year. During this time, detailed operational data on the performance of individual components, as well as the entire system, will be obtained. The overall schedule is shown in Figure 16-1.

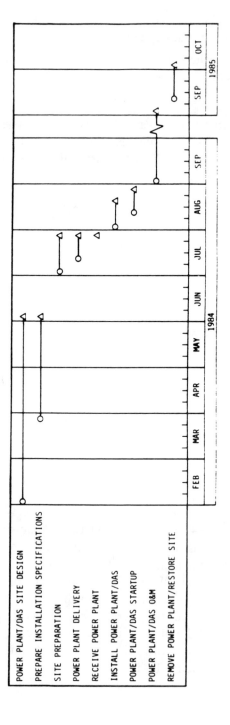

FIGURE 16-1. Wright-Patterson AFB Fuel Cell Project Schedule.

During field testing, routine and unscheduled maintenance activities will be performed by Wright-Patterson AFB personnel. Instrumentation will be periodically checked to ensure the accuracy of the test data. UTC will be notified in the event of an unscheduled shutdown of any power plant or incident of unusual significance. The goal is to operate the power plant for 8,000 hours.

Upon completion of the one-year field test, the power plant will be removed and returned to UTC, and the site will be restored to its original condition.

## SITE CHARACTERIZATION

The United States Air Force Museum is located on the Wright-Patterson AFB approximately 6 miles northeast of Dayton, Ohio. It is the oldest and largest military aviation museum in the world. The indoor museum viewing area (Building 489) is composed of approximately 130,000 square feet of exhibit area, a theater, gift shop, cateferia, administrative offices and small workship. More aircraft and missiles are located outside at an adjacent runway and at a temporary museum annex nearby.

Electricity is used to meet the museum's electric and thermal energy loads. Dayton Power & Light supplies 12 kV power to two 2000 kVa transformers at the museum. Space heating is provided to the exhibit area via twelve large air handing units with integral electric strip heaters. Two electric chillers (356 tons and 380 tons) provide chilled water to heat exchanger coils located in the air handling units. Domestic hot water is supplied to the cafeteria and restrooms by an electric hot water heater. The domestic hot water loads are minimal relative to the 40-kW fuel cell power plant heat production.

The 40-kW fuel cell power plant is an onsite energy system that uses pipeline gas for fuel to simultaneously generate AC electric power and recover thermal energy. The power plant is compact (6.5 ft. x 10.5 ft. x 5.3 ft.), quiet, low polluting, of modular design, and can operate unattended.

Figure 16-2 shows a simplified block diagram of the power plant's subsystems. Natural gas enters the fuel processor, mixes with steam and is catalytically converted into a hydrogen-rich gas. After cooling, the gas flows to the power section. The power section electrochem-

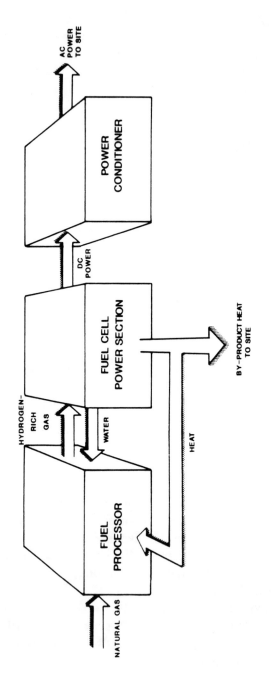

FIGURE 16-2.  Fuel Cell Power Plant.

ically consumes hydrogen from the gas and oxygen from the process air system, as it produces direct current. Water is produced as a by-product of the electro-chemical process.

The depleted fuel flows to the reformer where it is burned to produce the thermal energy required for the steam reforming process. Reformer burner exhaust combined with depleted air from the power section is sent to the heat exchangers to be cooled for heat and water recovery. The water is used for fuel processing needs.

A thermal management subsystem controls power section temperature by circulating water through the power section. Heat generated during the power production process is removed from the power section by changing the circulating water into a two-phase mixture of steam and water. This two-phase mixture flows through heat exchangers to provide thermal energy and power plant thermal control. Steam is separated for use in the fuel processor and the remaining water is recirculated through the coolant loop.

The power conditioner (inverter) converts the unregulated DC output from the power section into voltage regulated, current limited AC power. Output power from the inverter assembly is three-phase, four wire, 120/208 ± 5% VAC, 60 Hz. The inverter has a built-in fault clearing capability.

The fuel cell power plant has an electric load following capability and instantaneous load response through its normal power range of 0 to 40 kW(e). Electrical efficiency is approximately 35% when operating from half to full rated power; heat recovery varies from approximately 60,000 Btuh at half rated power to 150,000 Btuh at full power. Since heat recovery is approximately equal to electrical output, a user requiring between half and full electrical load, who can concurrently use all of the available by-product heat, will realize a 65-70% fuel efficiency. Total fuel utilization as a function of output power is shown in Figure 16-3.

The fuel cell provides low grade heat up to 180°F and high grade heat up to 275°F. This heat is normally supplied as heated water although small quantities of saturated steam can be produced. The 40-kW fuel cell generates utility quality power with low noise and a relatively low level of emissions.

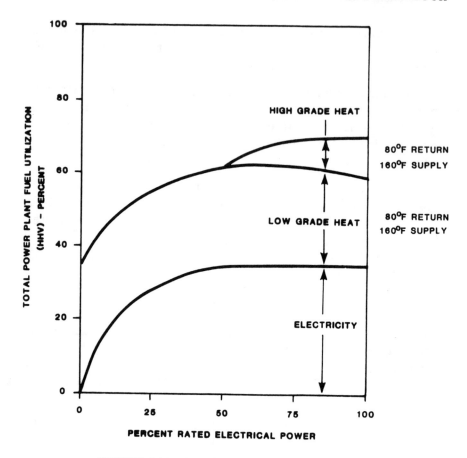

FIGURE 16-3.  Total Power Plant Fuel Utilization.

## SYSTEM DESIGN

The fuel cell provides output power at 120/208V, three-phase, four wire wye, however this amount does not satisfy the museum's total load. After investigating the existing electrical system, it appeared that the logical circuit to tie into appeared to be either an existing 225 kVA or a 150 kVA transformer.

However, these were both located in the basement of the facility. The only nearby circuit with sufficient capacity for the fuel cell power plant was a 480 V delta circuit. SAI chose to install a transformer after results of an economic analysis showed that running electrical lines to the basement would be more expensive. Figure 16-4 is a schematic of the electrical system.

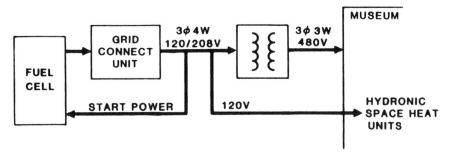

FIGURE 16-4. Electrical System Schematic.

### Electrical Gas System Design

The fuel cell stack produces DC power which is converted to 120/208V AC power in an internal inverter. To ensure compatibility with the utility grid system, the power first passes through the grid connect unit which contains protective equipment. Dayton Power & Light (DP&L) reviewed the electrical interconnection design and approved the safety devices. DP&L was extremely cooperative throughout the design and installation of the power plant.

The power is supplied to a 45 KVA step-up transformer to deliver 480V electricity to the museum. Additional or special metering was not required since the power produced would be consumed by the museum and no electricity would be fed back into the utility's lines. The protective features of the grid connect unit automatically isolate the fuel cell power plant in the event of a power outage. Since the power plant consumes 21 kW during start-up, the 45 KVA transformer operates as a step-down transformer at that time.

Since the Air Force Museum is an all-electric building, supplying natural gas to fuel the power plant presented a problem. The nearest natural gas pipeline was approximately 1700 feet away. Fortunately, a natural gas line was being considered for a planned extension of the museum. Therefore, it was decided to install a natural gas line that could meet the museum's future requirements and the needs of the fuel cell power plant.

**Thermal System Design**

The most critical factor in the overall compatibility of the installation was the thermal interface between the fuel cell and the museum. The only significant thermal load is space heating, which is provided by electric strip heaters in the fan coil units. These heaters are not compatible with the hot water heat recovery system of the fuel cell.

Air conditioning is supplied by a chilled water system which is drained during the winter months. The possibility of supplying hot water to an existing chilled water coil in one of the fan coil units was evaluated. The cooling coil would have been sufficient to reject all of the heat from the fuel cell; however, the concept was not used for the following reasons:

- The fan in the air handling unit does not operate continuously. The fan the the strip heaters are activated by a thermostat in the museum.

- The fan does not operate after normal museum hours.

- The fan motor is quite large and operating it only to reject heat from the fuel cell would be too expensive.

- Using the existing chilled water lines could present problems, e.g., water quality incompatibility, excessive corrosion, leaks, etc.

- During the spring and fall the museum could require both heating and cooling. Therefore, the fuel cell heating system could not be activated until air conditioning would no longer be required. This would seriously restrict the amount of heat that the fuel cell could supply.

Installing a separate, smaller forced air unit to reject the heat to the museum would avoid these pitfalls. Furthermore, the museum's large air volume could be used for thermal storage. At night, when the thermostats did not call for heat, the fuel cell system would continue to heat the air so that less heat would be required the next morning when the thermostats were reactivated.

The hydronic unit heater requires 160°F hot water at 6.25 gpm with 70°F entering air. To reject 150,000 Btuh from the fuel cell power plant, the temperature drop across the unit heater had to be

51°F. Figures 16-5 and 16-6 are thermal performance maps for the two fuel cell heat exchangers.

No off-the-shelf, reasonably priced hydronic unit heaters could meet these requirements. The only option was to reduce the requirements and either increase the air flow of the unit heater or increase its heat exchanger surface area. Although the latter option was more expensive, increasing the air flow by increasing the fan size would cause intolerable noise levels within the museum.

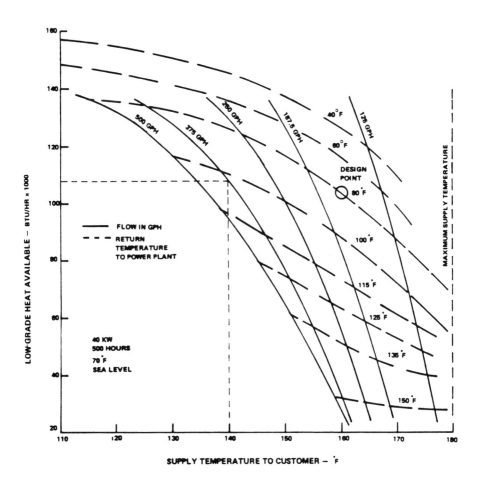

FIGURE 16-5. Low-Grade Heat Exchanger Performance Map.

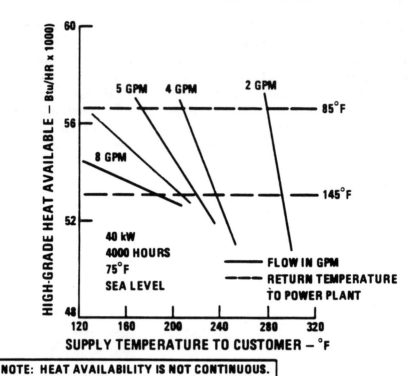

FIGURE 16-6. High-Grade Heat Exchanger Performance Map.

The heat exchanger surface area could be increased by installing multiple unit heaters. Due to technical constraints of commercially available hydronic unit heaters, the requirements were reduced to 140°F supply hot water, 6.25 gpm, and 70°F entering air. This reduced the maximum possible heat rejection from the fuel cell power plant to 130,000 Btuh with a temperature drop across the unit heaters of 40°F.

Also, the unit heaters increase the temperature of the entering air to 76°F. This small temperature increase produces heated air which would feel cool if blown directly on visitors at the museum. Therefore, care was taken to direct the flow of air away from visitors.

The fuel cell thermal operating characteristics are somewhat in-compatible with the hot water requirements of space heat hydronic unit heaters. The unit heaters are designed to operate on a low temperature differential (approximately 20°F or less), with a high temperature hot water supply (approximately 200°F), a relatively high water flow rate and to produce a high exit air temperature (approximately 110°F). The fuel cell, however, provides its optimum performance the cooler the return water which results in a large temperature differential. This is shown by the fuel cell power plant heat exchanger maps (Figures 16-5 and 16-6).

The high grade and low grade heat exchangers were installed in a series connection, because there were no apparent benefits to parallel connection. A parallel connection is applicable for two different hot water uses that have differing temperature requirements such as space heating and domestic hot water. A schematic of the thermal energy system is shown in Figure 16-7.

Although the overall heat production of the fuel cell power plant will not be completely used, the incompatibilities previously discussed will only reduce thermal performance by approximately 13% and should not affect the electrical performance.

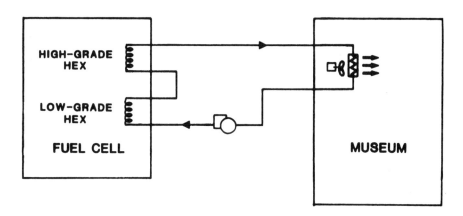

FIGURE 16-7. Thermal Energy System Schematic.

### Environmental Issues

Both the air and water emissions from the power plant are relatively benign. The exhaust emissions are less than 1/10 of the emissions from a natural gas-fueled steam power plant. The waste water is considered potable. Also the discharge of the water is very small, less than 1 gph during normal operating conditions. The noise levels are also relatively low at 60 dBA, fifteen feet horizontally from the fuel cell power plant.

These environmental considerations were discussed with the Wright-Patterson AFB Environmental Group, the Regional Air Pollution Control Agency, and the Ohio Environmental Protection Agency. The noise and air emissions were within acceptable ranges. The Ohio EPA requested that the waste water be drained to a sanitary sewer or dry well. Since the discharge was relatively small, the water was drained to an adjacent, rock-gravel area where it could evaporate.

## SITE DATA COLLECTION

The purpose of the data acquisition system (DAS) is to collect data vital to the operation and maintenance of the power plant and analyze this data to determine performance characteristics.

Figure 16-8, a simplified block diagram of the DAS which was installed at the Air Force Museum, shows the flow of data from electrical sensors (select load panel) and thermal sensors (water and gas flowmeters, resistance temperature devices) to the data logger and data recorder. The data logger, which is capable of monitoring up to 75 data points, scans each data point every 10 seconds. Average, maximum, and minimum values are recorded by the data recorder every half hour on magnetic tape cassettes.

To accurately measure the fuel cell's thermal output, the water flow and the inlet and outlet water temperatures of the fuel cell low and high grade heat exchangers are monitored by water flowmeters and resistance temperature devices (RTD's). The electric output downstream of the grid connect unit is measured using a select load panel and the gas consumption is measured with a gas meter equipped with an index transmitter. This information is collected and stored in the datalogger and recorder for further performance analysis. Outside ambient and indoor museum temperatures are also monitored.

To accurately analyze power plant operation and to detect potential problems, significant operating parameters such as controller logic outputs, output power characteristics and power plant system characteristics are recorded.

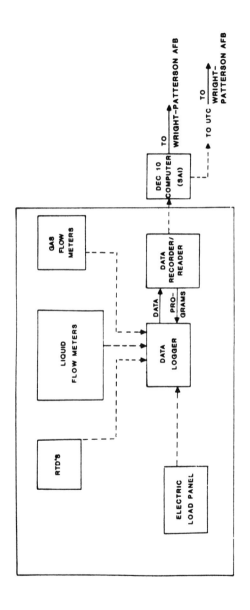

FIGURE 16-8. Data Acquisition System (DAS).

### Data Acquisition and Analysis

Every week, Wright-Patterson AFB personnel send a data tape to SAI for initial analysis. SAI's computer program analyzes the data, provides a statistical correlation of each parameter, calculates and provides power plant performance data and summarizes the performance. A plot of the fuel cell electrical output and the total fuel utilization is also generated.

Table 16-1 is an example of the type of power plant performance data gathered at the Wright-Patterson site. Every 4 hours data such as high and low grade heat exchanger efficiency, total efficiency, and the thermal capacity factor are calculated. Table 16-2 summarizes the power plant performance data over the time period of the tape.

Figure 16-9 is a plot of the fuel cell electrical output and the corresponding fuel utilization efficiency over the data collection period. A listing of the data used to generate the plot is also contained in the report.

Although the information in Tables 16-1 and 16-2 and Figure 16-9 is representative of data obtained from the Wright-Patterson site, the data is from another site in the GRI Onsite Fuel Cell Program. The energy plots and associated data provide a general overview of the power plant's operation at the site and serve as an indicator of potential problems and overall performance during the data time period.

### CONCLUSIONS

- The 40-kW fuel cell power plant which is operating in a cogeneration mode to supply electricity and space heat to the Air Force Museum is a demonstraton of the viability of fuel cells for the U.S. Air Force.

- Fuel cell power plant heat production and space heating applications are not completely compatible. However, the concept is workable with only a minor degradation in thermal performance and no effect on electrical performance.

- Analysis of the data and operating experience with the 40 kW fuel cell power plant at the museum site will provide important technical information for future fuel cell applications at Air Force bases.

TABLE 16-1. Power Plant Performance Data Extract.

| DATE | TIME | AVG KW | HIGH GRADE BTU/HR | HIGH GRADE T-IN | HIGH GRADE T-OUT | HIGH GRADE GPH | BTU/HR | LOW GRADE T-IN | LOW GRADE T-OUT | LOW GRADE GPH | GAS CFH | ELECT EFF(%) | TOTAL EFF(%) | THERM CAP(%) |
|---|---|---|---|---|---|---|---|---|---|---|---|---|---|---|
| 5/11/84 | 15:22 | 39.9 | 45415. | 147.8 | 158.9 | 490.5 | 89632. | 125.9 | 147.8 | 490.5 | 420. | 31. | 65. | 90. |
| 5/11/84 | 19:22 | 39.5 | 43611. | 116.2 | 127.0 | 484.8 | 149868. | 79.2 | 116.2 | 484.8 | 402. | 32. | 82. | 129. |
| 5/11/84 | 23:22 | 39.5 | 25771. | 144.4 | 150.9 | 478.2 | 89620. | 122.0 | 144.4 | 478.2 | 366. | 35. | 68. | 77. |
| 5/12/84 | 3:22 | 39.1 | 20983. | 150.3 | 165.3 | 393.3 | 66267. | 128.1 | 150.3 | 393.3 | 362. | 35. | 61. | 58. |
| 5/12/84 | 7:22 | 39.1 | 19120. | 150.6 | 169.8 | 346.4 | 81784. | 126.2 | 150.6 | 346.4 | 354. | 36. | 61. | 54. |
| 5/12/84 | 11:22 | 39.5 | 22293. | 119.9 | 137.1 | 360.6 | 121049. | 76.1 | 119.9 | 360.6 | 368. | 35. | 76. | 96. |
| 5/12/84 | 15:22 | 39.6 | 17128. | 145.6 | 168.5 | 294.2 | 79030. | 107.2 | 145.6 | 294.2 | 374. | 34. | 62. | 64. |
| 5/12/84 | 19:22 | 39.5 | 18707. | 116.4 | 131.2 | 370.5 | 125328. | 73.9 | 118.4 | 370.5 | 366. | 35. | 76. | 96. |
| 5/12/84 | 23:22 | 39.2 | 11796. | 147.6 | 165.2 | 333.8 | 64583. | 124.3 | 147.6 | 333.8 | 358. | 36. | 59. | 51. |

TABLE 16-2. Power Plant Data Summary for Periods of
Power Plant Operation Only.
(HHV = 1000.0 Btu/ft**3)

| | |
|---|---|
| 96.0 | Hours of data on cartridge of which: |
| 95.0 | hours are for fuel cell operation |
| 39.1 | kW, average power level during operation |
| 1142144.3 | Btu, total thermal energy from high grade hex |
| 7914357.7 | Btu, total thermal energy from low grade hex |
| 97.8 | % electric capacity factor |
| 31.8 | % thermal factor |
| 63.1 | % fuel utilization during operating period |

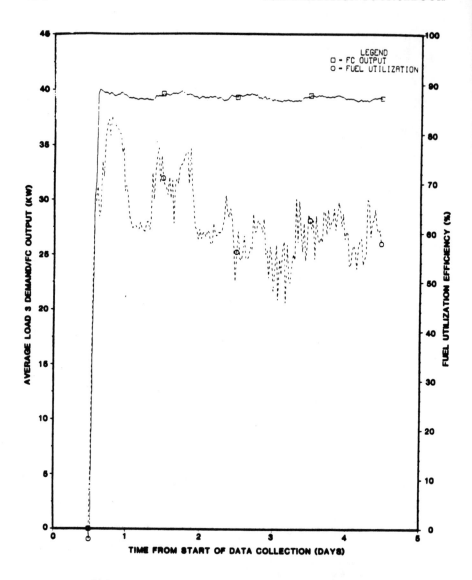

FIGURE 16-9.  Fuel Cell Power Plant Performance Plot

**REFERENCES**

Staniunas, J.W., *On-Site Fuel Cell Application Analysis*, AFWAL-TR-83-2014, February 1983.

Ferraro, V.D., and Arcari, R.N., "Site Selection for Onsite Fuel Cell Power Plant Testing," presented at the 19th IECEC Conference, August 1984.

Racine, W.C., et. al., "Characteristics of Candidate Sites Selected for Onsite Fuel Cell Power Plant Testing," presented at the 18th IECEC Conference, August 1983.

*Fuel Cell Power Plant Specification for Field Test Model*, FCS-1460, Revision C, United Technologies Corp., Power Systems Division, July 1983.

Allen, M., and Fellner, Lt. J.P., "Air Force Fuel Cell Program," June 1984.

# CHAPTER 17
# An Integrated Approach to Cogeneration Planning Using Renewable Energies

*R.N. Basu, L.L. Cogger*

Cogeneration in current engineering terminology has come to mean simultaneous generation of heat and electricity from the same power station. But there is another meaning of the word, i.e., coincidental and/or cooperative generation of power from different sources. Both of these aspects are covered in this chapter, which examines the options on renewable energy available to energy planners and gives examples of typical resource analysis. The system analysis for choice of the most suitable option as well as some of the important characteristics of renewable energy resources will be discussed.

## OPTIONS OF RENEWABLE ENERGY CONVERSION

Figure 17-1 shows the renewable energy sources that have been considered suitable for central-station generation of electricity. Renewable forms of nuclear and fusion energy have been excluded from this chart because the emphasis in this chapter is on alternative and non-polluting sources of power.

The chart is also biased towards the technologies of renewable energy application which are well-developed and available for use. During the last two decades, these technologies have gone through varying degrees of development. Although some applications have been halted in their progress because of operational and economic factors, this does not preclude the possibility of further developments in the future. Reviews describing the current technology of various modes of renewable energy conversion show that considerable progress has been made.[1,2]

The economic potential of a particular mode of renewable energy conversion is quite different in different places. What is uneconomic

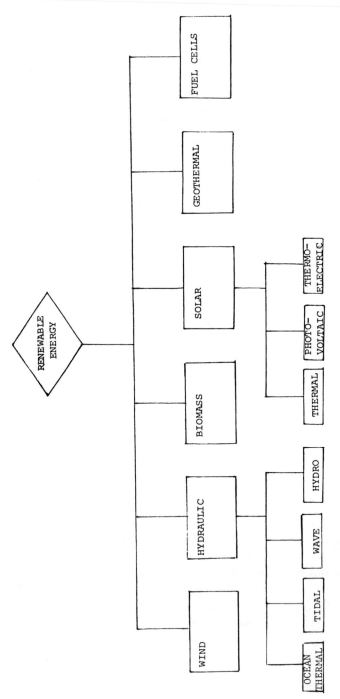

**FIGURE 17-1. Currently Available Major Renewable Energy Options for Cogeneration.**

in Alberta (Canada) is not necessarily so in Assam (India) or in Arizona (USA).

Solar photovoltaic conversion is at present the most promising technology for the future. A one-MW facility is already in operation in Hesperia, USA, and others are projected for the Middle East countries. The success of the amorphous method of solar cell production is promising to reduce the costs to a level where solar photovoltaic electricity would compete with the cost of conventional small thermal and other generating modes.

The potential of biogas is also high on a worldwide scale. Here is a fuel of high calorific value produced from waste materials which have been traditionally considered to be a nuisance and costly to dispose of. Now farmers all over the world are cogenerating, i.e., generating heat and electricity by burning methane obtained from fermentation of piles of refuse, cattle dung, chicken manure, etc.[3] The material left after the extraction of methane from a digester becomes a valuable fertilizer.

Electricity generation from wind is also developing fast after teething troubles. According to an estimate, by the year 2000, as much as 10% of California's electrical energy could be produced from renewable wind resources.

## 9-F MODEL FOR SYSTEM ANALYSIS

A block diagram showing the structure of the system analysis methodology and of the steps in the design of a cogeneration system, is presented in Figure 17-2. The process can be complex because of the possible number of choices available.

The initial problem definition would include a demand and load analysis. This is to be followed by resource analysis. The latter would include a study of the critical characteristics of the resources. In the diagram the resources are characterized by either the form of natural energy or the fuel. The critical properties are summarized under the headings of fixity, friendliness, frugality, functionality and future prospects. These properties decide the short and long term success of cogeneration with any form of energy.

Once these analyses are completed, and their results are available, the design of the cogeneration system can proceed. Primary and

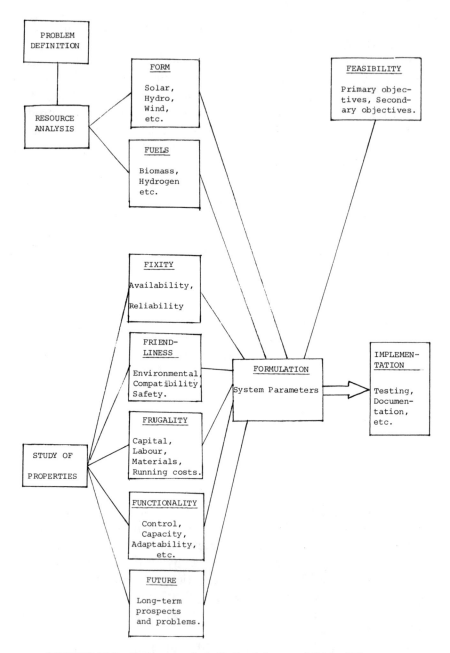

**FIGURE 17-2. System Analysis Methodology and Critical Elements in Cogeneration Planning — 9-F Model.**

secondary objectives are set for the design, and the feasibility of meeting these objectives is studied.

After the feasibility study is completed, the final parameters of the system are formulated. The system can now be implemented.

Implementation includes testing of the system and preparing the final documentation.

Software packages developed in the past can be adapted for performing different tasks in cogeneration planning. Application of programs at Rensselaer Polytechnic Institute, i.e., OPLAN (optimal generation planning program package), PREP (generation planning data base), PROFIL (processor of hourly loads) etc. have been reported[4] for generation planning programs. The system model has also been used for determining "the feasibility of novel technologies such as solar electric generators, wind turbines and biomass as a fuel."

A sequential multiple objective analysis method reported by the same group,[5] is a useful technique for developing alternative strategies with overlapping objectives. In this method, an optimum strategy is first developed based on the primary objective (e.g., minimum cost). A modified strategy is then obtained based on a secondary objective (e.g., energy conservation) with a permissible deviation from the primary optimum in order to obtain the secondary objective. The underlying mathematical technique of linear programming can be applied to the mathematical models for comparing different combinations of renewable resources for cogeneration.

The artifice of grouping the nine critical elements/steps under titles starting with the letter F provides coherence in coding and storing the data in correspondingly named files. Accordingly, the model has been called the 9-F model.

## DEMAND AND RESOURCE ANALYSIS

Given a situation where a demand for power has to be met by a combination of renewable energy resources, the starting step is a demand and resource analysis.

The energy demand is variable with time and so are most of the renewable resources in their natural forms. The task is to match the supply to the demand. Whether the demand is existing or anticipated, aggregate power consumption would be plotted to show the maximum requirement over any period of time. The capacity of the energy supply would be selected on this basis.

In resource analysis, first the physical potentials of renewable energy in the region is surveyed. A typical report on the physical potential done on Kenya and Tanzania by the German Appropriate Technology Exchange (GATE)[6] is shown in Table 17-1. The potential survey can then be followed by a more detailed system analysis using the 9-F model of Figure 17-2, to decide on a suitable cogeneration plan using chosen resources.

Table 17-2 shows estimates of the total power needs of the United States that will be supplied by solar energy in the years 1985, 2000, and 2020. These estimates were made by the U.S. Energy Research and Development Administration (ERDA).[7]

The order of growth of different options of renewable energy in these estimates differ from that of global developments so far. However, the general trend for the year 2000 onwards which expects photovoltaic solar electric capacity to exceed other forms of generation appears to be correct. The development of solar thermal electricity seems to be slowing down in many countries including Japan which used to be a strong advocate. The development of ocean thermal electricity is also growing much slower than estimated.

In farms and rural areas, energy demand patterns are different from those in metropolitan areas. The per capita consumption is smaller in the rural areas, particularly in the developing countries.

However, the energy demand of commercial farming has been rising steeply. The average annual use of electrical energy on United States farms in 1973 was 12,588 KWh.[8] The corresponding figure in other countries was lower, probably with the exception of Canada. The vast discrepancy in energy consumption between different parts of the world is apparent from Table 17-3 published by the Food and Agriculture Organization.

Where the energy demand is moderate, an independent power supply based on cogeneration by solar thermal, biomass and photovoltaic conversion modes is highly feasible and sometimes preferable from an energy conservation point of view.

Paradoxically, a farm has within itself resources for producing alternate energy, i.e., space for collecting solar energy, animal and vegetable matter for generation of biogas, etc.

## TABLE 17-1. Physical Potentials of Renewable Energy

Group II: *Kenya, Tanzania*

*Solar Energy:* Radiation and sunshine hours varying strongly by regions due to seasons and altitude. In total good and fairly good potentials, in the North of Kenya (Lodwar) very good.

Annual means: 4.5 - 6.2 kWh/m$^2$/day   Fluctuations of monthly
5.5 - 9 sunshine   means in relation to annu-
hours   al mean of about 25%

*Wind Energy:* In coastal regions and North Kenya good potentials due to low fluctuations during the year. Moderate trade wind with low storm probability (v= 10m/s)

Annual means: 4 - 5 m/s in favorable regions
2 - 3 m/s country average

*Small hydropower:* Regionally limited good to fairly good potentials in the mountain regions. Limited fluctuations during the year. Some potentials within already existing irrigation and drinking water systems. The quantification of possible sites has been limited up to now to certain regions (West Kenya, Tana valley).

Biomass: Very good and good potentials of agricultural waste (straw, sisal residues, coffee pulps, peanut husks, banana-peels). Regionally varying forest-potentials depending on the intensity of agriculture. Good potential of grass and reed near the great lakes. Animal excrements occur decentralized and regionally limited. Some farms with intensive animal husbandry offer good potentials.

Quantitative assessment of agricultural waste:   Kenya   Tanzania

Million tons dry residue per year   n.a.   9

TABLE 17-2. Estimates of the Heat, Electric Power, and Fuels to be Supplied by Solar Energy in the United States, as Projected by ERDA.

| Solar technology | 1985 | | 2000 | | 2020 | |
|---|---|---|---|---|---|---|
| Direct thermal application (in units of $10^{15}$ Btu = 1 Q per year)[b] | | | | | | |
| Heating and cooling | 0.15 | Q | 2.0 | Q | 15 | Q |
| Agricultural applications | 0.03 | | 0.6 | | 3 | |
| Industrial applications | 0.02 | | 0.4 | | 2 | |
| Total | 0.2 | Q | 3 | Q | 20 | Q |
| Solar electric capacity (in units of $10^9$ W = 1 GWe)[c] | | | | | | |
| Wind | 10 | GWe | 20 | GWe | 60 | GWe |
| Photovoltaic | 0.1 | | 30 | | 80 | |
| Solar thermal | 0.05 | | 20 | | 70 | |
| Ocean thermal | 0.1 | | 10 | | 40 | |
| Total | 1.3 | GWe | 80 | GWe | 250 | GWe |
| Equivalent fuel energy | 0.07 | Q | 5 | Q | 15 | Q |
| Fuels from biomass | 0.5 | Q | 3 | Q | 10 | Q |
| Total solar energy | ∿1 | Q | ∿10 | Q | ∿45 | Q |
| Projected U.S. energy demand | 100 | Q | 150 | Q | 180 | Q |

[b]One Q represents the approximate energy consumption of the state of Iowa in 1973.
[c]One GWe is one gigawatt of electrical power.

TABLE 17-3. Estimated Commercial Energy Consumption of Agriculture, 1972/73.

| | Agricultural energy consumption, percentage share | Energy consumption per capita of an agricultural worker (in $10^9$ joules) |
|---|---|---|
| Developed Countries | 3.4 | 107.8 |
| Western Europe | 4.9 | 82.4 |
| U.S.A. | 2.8 | 555.8 |
| Developing Countries | 4.8 | 2.2 |
| Latin America | 3.8 | 8.6 |
| Africa | 4.5 | 0.8 |
| Asia | 5.3 | 1.4 |

Source: FAO 1979

## FORMS OF ENERGY AND FUELS

The present status of different forms of renewable energy has already been commented upon in the previous sections. More detailed discussion is beyond the scope of this chapter.

Hydrogen has the potential of becoming one of the unlimited fuels of the future. But bulk storage, safety and reliable combustion devices are still under study.

In their Plan 2000, Brazil has included a 50 MW thermal plant with wood fuel and the use of biodigesters for electricity production from sugar cane bagasse. The future prospects of these methods of cogeneration would depend on planned reforestation and efficiency of crop production and of the fermentation process. These aspects are being studied in Canada and elsewhere.

## FIXITY OF RENEWABLE POWER SOURCES

One major shortcoming of many of the renewable energy sources is their lack of fixity, i.e., variable availability. Hence a degree of storage has to be provided for solar, wind and tidal schemes to compensate for the stochastic nature of the energy source.

A.K. David has presented a method for deriving mathematical models[9] for representing sources of this type, without storage, that can be incorporated in well-established methods of power system reliability evaluation. Power generation is written as a function of intrinsic variables which describe the state of the relevant elements of the physical environment. For a 3-parameter solar powered electric generator system, for example, power generation is modelled by a nonlinear equation and three different types of probability density functions are chosen for the three parameters. The power output Z is related to the intrinsic parameters by a function of the type:

$$Z = f_1 (x_1) f_2 (x_2) / f_3 (x_3),$$

where x = deviation of solar inclination from the vertical (i.e., x = 0 when overhead)

$x_2$ = degree of cloud cover on a percentage scale from 0 to 100%

$x_3$ = ambient temp on an arbitrary scale

After obtaining a prior probability distribution derived from experience and conditioning observations, available power probabilities are calculated. The probability results may be employed for computing the reliability indices of systems while knowledge of transition rates facilitates frequency and duration analyses, as well as decisions regarding spinning reserve assessment.

The other aspect of Fixity is reliability, i.e., immunity from failures and operating life. Reliability is usually quantified as mean time between failures. It is determined statistically by field testing or simulated life testing. As many of the renewable power sources are still relatively new, there is a lack of reliability data on some of the devices used.

## FRIENDLINESS OF RENEWABLE POWER SOURCES

This has several aspects: (1) Environmental impact; (2) Safety aspects; and (3) Compatibility with other energy modes and other activities. Friendliness to environment is a positive factor in favour of the renewable sources in general which has led to their public support worldwide.

However, compatibility with existing power resources, i.e., conventional power, can be an important factor in deciding the choice. Many of the renewable power modes are relatively new in the public domain; some aspects have not yet been fully studied for each mode. The "Solar Heating Materials Handbook"[10] shows the multiplicity of environmental and safety considerations involved in the selection process of solar heating materials alone.

## FRUGALITY OF POWER SOURCES

A complete assessment of the frugality, i.e., economic performance, of different types of power generating systems is a complex task. To be truly comprehensive, it has to take into consideration not only the first costs, i.e., capital, materials and labour, and the running costs, i.e., fuel, maintenance, etc. but also life cycle costs of pollution, etc.

The cost of repairing the damages of acid rain and of compensating miners suffering from silicosis are examples of the latter type of costs associated with coal-fired power generation. The cost of storage of nuclear waste is the corresponding example for nuclear power generation.

The fuel cost in the renewable energy options discussed in this chapter range from zero in case of solar and hydro to moderately high in case of methanol. The plant and installation costs for different options are also widely different.

Solar energy is generally not considered to be cost-effective, for supplying full load requirements involving thermal and mechanical loads for long periods of time. The latter would have to include sunless periods and periods of maximum demand.

Consequently, the solar system would need to incorporate large storage capacity which is expensive.

A better solution would be for solar power plants to provide intermediate and peak loads while the base load is supplied by some other form of energy, e.g., hydro or geothermal or conventional fuel-fired.

An equation for calculating the approximate savings achieved by the installation of a solar system for supplying the part of the total land would be as follows:

$$S = C_f - (C_s - C_N) - (E_s - E_N), \text{ where}$$

$S$ = Net savings during life cycle

$C_f$ = Conventional fuel costs saved during life cycle

$C_s$ = Installed cost of solar system

$C_N$ = Installed cost of equivalent non-solar system in case this was applicable

$E_s$ = Running expenses of solar system during life cycle

$E_N$ = Running expenses of non-solar system excluding fuel during life cycle

## FUNCTIONALITY OF POWER SOURCES

Any system supplying heat and electricity for practical applications must meet certain specifications laid down for the purpose. These include parameters like temperature, voltage, current, frequency, etc.

The parameters of natural energy, e.g., solar heat, wind, tides, etc. are often not ideal for direct utilization. Hence special equipment has to be interposed for transformation and control of the natural energy to meet functional requirements.

For example, the low-voltage direct current obtained from solar photovoltaic cells has to be inverted and stepped up to be able to run conventional industrial equipment. These interfacing requirements are not over-complicated for the ultimate energy options discussed in this paper.

The existing technology of electronic sensors, transducers and control apparatus have been adapted successfully for the control of equipment designed for exploitation of renewable energies. Although one is dealing with low temperatures (except in geothermal) and low pressure, special problems are posed by corrosion (ocean) and turbulence (wind and tides) in designing satisfactory mechanical equipment.

Microcomputer control is predestined to be used for regulation of all the generation and cogeneration systems.

## FUTURE OF POWER SOURCES

When designing a power generating system, one has always to look towards the future. This is particularly true for large generating systems like thermal or nuclear power stations because of the long lead time for the stations to come on stream. But it is true for renewable energies as well.

In recent years, sometimes called the age of uncertainty, energy planners have run into some difficulty due to their load forecasts becoming upset due to sudden unpredictable changes in the social, financial, political or environmental conditions of their locale. As a rule, energy demand forecasts of Western planners have been found to be overoptimistic while those in the developing countries have been pessimistic. Consequently, there are a few mothballed nuclear and thermal plants in Europe, U.S.A. and Canada while Asia and Africa continue to suffer from power blackouts and brownouts.

However, error in forecasting is not the only factor responsible for this situation. Lack of maintenance or choice of unsuitable form of design are often contributary factors. Many energy devices have failed prematurely because of lack of adequate provision for planned maintenance, aging of materials, etc.

When planning a cogeneration system for a lifetime of 25 years, one ought to prepare for anticipated changes during this quarter century of rapid innovation and societal transitions.

## SOLAR PHOTOVOLTAIC AND THERMOELECTRIC MODULES

From the options available for cogeneration of heat and electricity from renewable energy resources, the promise of the above two modes, which are at present underdeveloped, is worth  mentioning. Both these modes offer direct conversion of solar photons into electricity. They are simple, silent, solid-state, with no moving parts. They are highly favourable from the criteria of fixity, friendliness, functionality, and fuel-independence and have good future prospects. But they are considered too costly in their present form.

Solar photovoltaic modules are becoming cheaper due to the improvements in the amorphous deposition of Si and other active materials. Solar thermoelectric modules are at present still in the research stage. Cogeneration by a combination of these two harmonious modes has great appeal because it offers the potential for much better utilization of the solar energy spectrum.

## A HYBRID WIDEBAND SOLAR COLLECTOR

Solar collectors obtain high-grade heat by using concentration and tracking methods. This is the basis of solar thermodynamic generation of electricity using heat engines or steam turbines. The heat outputs of all solar collectors are dependent on the collector area, i.e., space.

The electrical output of solar photovoltaic modules and of solar thermoelectric modules is proportional to the area of active surface which is again related to space.

In the hybrid collector, an attempt is made to minimize the space and hardware requirements of the three modes jointly by using a multi-tier approach within one framework. Thereby, the solar collector becomes a true cogeneration device.

One simple form of the design is shown in Figure 17-3. It is a double-parabolic collector with elliptical-tube heat-receptors (marked

AT) in its focal planes. The tube surface running at temperatures above 200°C would generate steam within, which is carried by a pulsed pumping method down to a storage device.

Underneath the receptor and in intimate contact with it is the hot junction of a thermoelectric generator (marked TE), the cold junction of which is in contact with another pipe (at A, A') below, carrying a cooling medium, e.g., ethylene glycol. This medium becomes the conveyor of low-grade heat. The thermoelectric generator would generate electricity due to the thermal excitation.

Electricity is also generated simultaneously by solar photovoltaic modules (marked FP) located at the two edges of the collector box frame. These are pivoted so as to be able to be turned towards the sun at fixed intervals. The backs of the SPV modules can be water-cooled. The cooling water would convey low-grade heat.

Such a design can provide relatively high conversion of solar energy into heat and electricity per square foot of projected area under the sun. The performance of the hybrid collector can be predicted by using a modified program based on Solar C[11]. The flow chart for such a program is given in Figure 17-4.

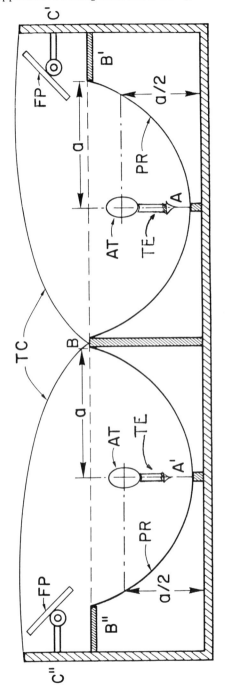

FIGURE 17-3. Cross-Section View of Double-Parabolic Solar Collector.

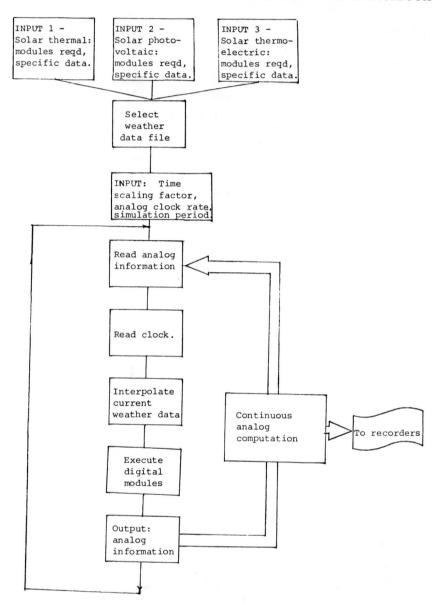

FIGURE 17-4. Simplified Simulation Flow Chart by Solar C
for Hybrid Solar Collector.

## REFERENCES

1 R.H. Taylor, *Alternative Energy Sources for the Centralized Generation of Electricity*, Heyden & Son, 1983.

2 Barb K. Harrah, *Alternate Sources of Energy*, Scarecrow Press, 1978.

3 T.J. Siebenmorgen et al., "Cogeneration of Electricity and Hot Water," Fact Sheet #7, Institute of Agriculture and Natural Resources.

4 Martin Becker et al., "Advanced Techniques for Electric Generation Planning," *Canadian Communications and Power Conference Proceedings 1978*, IEEE Cat. No. 78 CH 13730 REG 7.

5 William Rutz, et al., "Sequential Objective Analysis for Generation Planning using Linear Programming Optimization Techniques," ibid.

6 C.P. Zeitinger, et al., "Analysis of the Planning Phase of the Special Energy Programme (SEP), *Gate Report*, August 1982.

7 *National Solar Energy Research, Development and Demonstration Program, Definition Report*, E.R.D.A., U.S. Govt. Printing Office, Washington, D.C., 1975.

8 E.T. Wolford, "Electricity in Agriculture," *Standard Handbook for Electrical Engineers*, 21-19, McGraw Hill, 1978.

9 A.K. David, "Availability Modelling of Stochastic Power Sources," *IEEE PROC.*, Vol. 129, Pt. C. No. 6, Nov. 1982.

10 U.S. Department of Energy, Pub. No. DOE/TIC-11374, *Solar Heating Materials Handbook*.

11 National Research Council Report No. SIM-4, "SOLAR-C: Computer Program for the Design of Solar Energy Systems."

# CHAPTER 18

# A Case Study of Cogeneration at a Mental Health Center

*S. Tuma, A. Khan, D. Jones*

Studies have shown that annual utility costs in buildings owned and operated by the State of Illinois have risen from $91 million (FY81) to $113 million currently and are expected to be $135 million at the end of FY85. These costs can only be covered by scarce tax dollars. The Illinois Department of Energy and Natural Resources organized and now operates an Illinois Audit and Local Energy Awareness Staff Training Program called the State Building Energy Program (SBEP) to help State building managers control these rapidly escalating expenses.

SBEP audit recommendations highlight areas within building systems where changes in normal operating and maintenance procedures can result in significant energy cost avoidance and increased control over utility expenditures. Audit results are then used to achieve the program's objectives including local staff energy awareness training and the forming of a follow-up program designed to ease implementation of the necessary local procedure changes.

Continued communications and local staff energy awareness training becomes increasingly important in this approach since most energy cost control suggestions mean changes to existing facility operational procedures as practiced day to day by local staff.

The cogeneration equipment discussed in this chapter gets an exceptionally high level of competent maintenance to ensure its reliable operation and useful life. SBEP supervision was concerned that program objectives were not adequate to properly address the operational requirements of facilities with such capabilities.

This case history grew out of concern that the benefits, costs, and maintenance need levels of such existing cogeneration installations in State buildings were not fully defined and appreciated and that further study was needed to understand their overall significance in terms of SBEP activity.

This study and analysis of the energy generation and cost characteristics of the Jacksonville Mental Health and Development Center power generating plant is based on data for February 1982–January 1983. The objective of this study is to broadly familiarize SBEP personnel with the aggregated costs and operational parameters of in-house power generation as it is presently being carried out at some Illinois facilities. It can also provide a preliminary data base to explore the potential for development of cost-effective power cogeneration projects at other state facilities as part of their energy and energy cost saving strategy.

It should be kept in mind that if results of this study are to be applied in other State facilities they will need to be modified considerably according to each local situation. Some of the significant factors affecting the applicability of these results are daily and seasonal electrical use patterns and the quantities consumed, fuel type used and cost, and supply reliability and local environmental considerations. Other factors affecting study results are the utility company's alternative electrical supply rate structure and terms of any power buy back contracts State generating facilities may enter into.

**South-Facing View of Primary 2,000 kW Cogenerating Plant. The two 750 kW Backup Units are Located to the Rear.**

## SUMMARY

During the year February 1982-January 1983, the Jacksonville MHD Center registered average high power demands of 920 kW in spring and a 1,160 kW high during the rest of the year. The plant produced 6-7 million kWh of electricity (including 0.3 million kWh used internally by power generating equipment) at an annual operational cost of $195,430 or 3.10¢/kWh.

In addition, the waste heat from the plant served to produce 2.1 million ton hours of cooling. Had the power been purchased it would have cost the facility $292,440. The in-house power generation therefore saved $97,000 which is 33% of the power purchase cost.

However, if the capitalized equipment cost is also taken into account, the cost of in-house generated electricity rises to 5.82¢/kWh which is 27% in excess of 4.57¢/kWh, the cost if purchased from the local utility company. The overall 20% efficiency of the plant compares favorably with the efficiency range of 7%-22% which is typical for similar cogeneration systems ranging in size from ½ to 7½ megawatts (MW).

## DESCRIPTION

Jacksonville MHD Center meets its entire electric power requirements from its 2 MW generator (1,700 kW output and built in 1947). It is powered by a condensing type single-stage steam turbine. The condenser is cooled by a standard cooling tower using two 40 Hp fans and a 40 Hp circulating pump. The admission steam is at 250 psi-400°F and the outlet hot well temperature varies from 110°F to 95°F. A part of the throttle throughput of steam is tapped ahead of the condenser at somewhat higher pressure. The 'extracted' steam at a pressure of 6 psi is used for heating both the domestic hot water, condensate return and for running the absorption chillers (2 - 400 ton and 2 - 200 ton units) during summer. In addition, the generation power plant has two more standby noncondensing type turbines each with 750 kW generating capacity.

Steam at 250 psi is generated by two coal fired water tube boilers each of 40,000 lb/hr. generating capacity. Two more similar boilers with capabilities of 40,000 lb/hr. and 45,000 lb/hr. are standby.

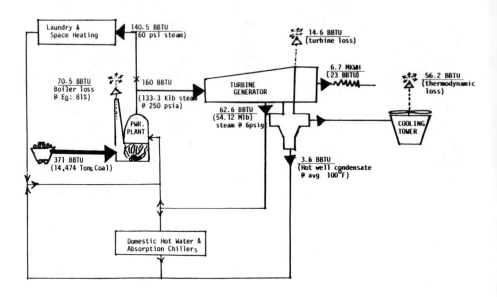

* Figures relate to period Feb. 82 - Jan. 83
@ Prefixes before units indicate as:

    K = Thousand
    M = Million
    B = Billion

**Energy flow diagram of Jacksonville, Illinois,
Mental Health & Development Center power plant.**

 

The cogenerating portion of the plant is monitored hourly by maintenance engineers who dedicate a total of 160 manhours per month (about 5 manhours per engineer per week) to the proper operation of the equipment. Their time is spent measuring 21 operating parameters each hour and monitoring proper governor speed control functioning. In addition they run water tests, conduct periodic blowdown and change or clean ejectors if vacuum measurements warrant it.

**Power Plant Building Showing Some of the
Cogenerating Equipment Maintenance Crew.**

## POWER DEMAND PROFILE

During the year February 1982-January 1983 the power plant produced 6,662,500 kWh including 305,400 kWh used by the electricity generating plant for internal consumption. The low in the daily demand cycle ranges from 600 kW to 650 kW and it occurs between 9 p.m. and 5 a.m. Daily demand cycles show a rather rapid rise from 9 a.m. peaking at about 11 a.m. The internal load peaks of the power plant range over 150-180 kW. The monthly demand highs registered during the year in question were as follows:

| April 1982 | 850 kW |
|---|---|
| May 1982 | 950 kW |
| June 1982 | 950 kW |

(a) Average spring high = 917 kW

| July 1982 | 1,250 kW |
|---|---|
| August 1982 | 1,150 kW |
| September 1982 | 1,100 kW |
| October 1982 | 1,250 kW |
| November 1982 | 1,100 kW |
| December 1982 | 1,000 kW |
| January 1982 | 1,150 kW |
| February 1982 | 1,250 kW |
| March 1982 | 1,200 kW |

(b) Average high for the rest
of the year = 1,161 kW

## ANALYSIS & RESULTS

The analysis is based on data shown in Appendix A and the assumptions made have been stated in Appendix B which also includes the calculations and the definitions of the various terms used below. In the following, numbers in brackets refer to items in Appendix B.

(a) About 53% of the total steam generated by the boiler system is diverted to the turbine but 43% of this turbine heat input is recycled back through the condensate return and the extraction loop. So, in effect, 30% of coal consumed in the plant with a cost of $172,560/yr.[1] goes toward net electric generation output of 6,357,119 kWh/yr.[5b]

(b) Extraction ratio based on nominal (unadjusted for heat reclaimed) throttle flow is 41%. The electric output to thermal input ratio based on nominal throttle flow is 50 kWh produced per 1,000 pounds of steam. On the basis of throttle flow adjusted for regeneration (internal waste heat reclamation) loop the electric output to thermal input ratio is 88 kWh produced per 1,000 lb. of steam. It may be noted for comparison that for similar cogeneration plants in sizes ranging from ½ MW to 7½ MW the typical ratio range is from 20 to 60 kWh per 1,000 lb. of steam.[2]

(c) Based on the annual data the average operating parameters of the generating power system are as under [3]:

    (i)     Nominal Steam Rate (NSR): 20 lb/kWh

    (ii)    Theoretical Steam Rate (TSR): 7 lb/kWh

    (iii)   Actual Steam Rate (ASR): 11 lb/kWh

    (iv)   Turbine Efficiency ($E_T$): 62%

    (v)    Boiler Efficiency ($E_B$): 81%

    (vi)   Thermodynamic Efficiency ($E_R$): 41%

    (vii)  Overall Efficiency ($E_O$): 20%

It may be mentioned for a rough comparision that overall efficiency (Eo) in conversion from fuel BTU to electric BTU of cogeneration plants varying in size from ½ to 7½ MW typically falls within the range of 7% to 22%.

(d) Steam requirements to satisfy load cycle phases:[4]

    (i)     Average electrical demand output (761 kW) = 15,220 lb/hr (38% of the single full boiler capacity)

    (ii)    Base demand (625 kW) = 12,500 lb/hr (31% of the single full boiler capacity)

    (iii)   Spring season peak demand (917 kW) = 18,340 lb/hr (46% of full capacity of one boiler)

    (iv)   Annual non-spring season peak demand (1,161 kW) = 23,220 lb/hr (58% of full capacity of one boiler)

    (v)    Average internal (generating equipment) peak demand of power plant (165 kW) = 3,330 lb/hr (8% of full capacity of one boiler)

(e) Generating Costs[6]

| Item Cost | Cost ($) | Unit Cost ¢/kWh | % Contribution of Cost Item to Operational Cost |
|---|---|---|---|
| (i) Fuel Cost only | $178,746 | 2.7926 | 91.2% |
| (ii) Maintenance | 3,624 | 0.0566 | 1.9% |
| (iii) Repair & part replacement | 2,000 | 0.0312 | 1.1% |
| (iv) Makeup water & water treatment | 11,062 | 0.1728 | 5.8% |
| (a) Total operational cost (51.7% of total cost) | $195,432 | 3.0536 | 100.0% |
| (b) Capital cost (48.3% of total cost) | 176,734 | 2.7614 | |
| (c) TOTAL COST | $372.166 | 5.8151 | |

The operational unit cost of 3.10¢/kWh produced by this plant compares favorably with the price of 4.47¢/kWh which the facility would be paying if it bought its power from the local utility company.[7f] By generating its own power the facility is effecting an annual savings of $97,000, which is 33% of the costs the facility would incur it if purchased electricity.[8a]

For a new plant, however, an additional cost of 2.76¢/kWh would be incurred because of capital costs of a new turbine installation.[6b] This would raise the unit cost to 5.82¢/kWh. In such a case the idea of in-house electrical generation is not a good one since there is an annual loss of $79.730.[8b]

## CONCLUSIONS

● The Jacksonville Mental Health and Development Center power generating plant is being maintained at the high performance level expected of systems of similar size and type.

• At the annual operating cost of $195,430 the plant produced a total equivalent of 8.5 million kWh which, if the facility had to purchase, would have cost $292,440. The in-house power generation, therefore, effected a saving of $97,000/yr. which is 33% of the cost if purchased.

• If a new installation of an electrical co-generating system of the type and size as the one at Jacksonville MHD Center is installed, annualized capital cost, when added to operational costs, would raise the unit costs of cogenerated power by 27% above the price charged by the local utility company.

• Although it will pay economically to continue operation of the Jacksonville Plant, installation of a similar new cogeneration system would not be cost effective elsewhere in Illinois.

## Appendix A
## Generating Power Plant Data of Jacksonville MHD Center
### February 1982-January 1983

Fuel value of coal = 12,810 BTU/lb

Total cost consumption = 14,474 tons

Cost of coal = $39.40/ton

Total steam production = 249,688,000 lb

Total steam used by turbine = 133,254,000 lb

Total steam to extraction = 54,120,000 lb

Electric energy produced = 6,662,500 kWh

Cost of water treatment = $3,800

Cost of maintenance based on 20 hrs/month = $3,624

Cost of maintenance parts = $2,000

Cost of makeup water to steam generating cooling tower = $7,262

Steam pressure upstream of turbine = 250 psig

Steam pressure at extraction = 6 psig

Steam pressure for the main building space heating = 60 psig

Hot well temperature = 95°-110°F

## Appendix B
## Calculations

1. Energy Inputs for Power Generation

   (i)  % steam generated to turbine =

   $$\frac{133.254 \times 10^6 \text{ lb/yr}}{249.688 \times 10^6 \text{ lb/yr}} \times 100 =$$

   53.37%

   Gross steam input @ 250 psia:

   $133.254 \times 10^6 \times 1201.1 =$
   $160.1 \times 10^9$ BTU/yr.

   Steam extracted @ 6 psig:

   $54.12 \times 10^6 \times 1,157.1 =$
   $62.6 \times 10^9$ BTU/yr.

   Return Condensate @ an assumed average
   temperature 100°F and makeup at 55°F.

   $(133.254 - 54.12) \times 10^6 \times (100 - 55)$
   $7.9 \times 10^9$ BTU/yr

   (ii) Net heat input for power generation
   (Net steam input to turbine)

   $(160.1 - 62.6 - 0.85 \times 3.5) \times 10^9$ Btu/yr.

   $\underline{90.79 \times 10^9 \text{ BTU/yr}}$

   (iii) Percentage heat input recycled in regenerative loop =

   $$\frac{(62.6 + 3.5 \times 7.9) \times 10^9}{160.1 \times 10^9} \times 100 =$$

   41.29%

   (iv) Coal used in power generation:

   $$\frac{94 \times 10^9}{249.688 \times 10^6 \times 1201.7}$$

   x 14.474 ton/yr = 4,536.7 ton/yr
   (31.34% of total plant consumption)

(v)  Cost of coal in power generation:

4,536.7 ton/yr x $39.4/ton =

$178,746/yr

(vi) Unit cost of coal energy:

$$\$39.9/\text{ton} \times \frac{\text{ton}}{2000\ \text{lb}} \times$$

$$\frac{\text{lb}}{12,810\text{BTU}} \times \frac{10^6\ \text{BTU}}{\text{MBTU}} =$$

$1.5379/MBTU (million BTU)


2.   (i)  Nominal turbine steam throughput for power generation:

$$\frac{133.254 \times 10^6}{249.688 \times 10^6} \times 100 =$$

53% of total plant production


(ii) Effective steam throughput for power generation adjusted for regeneration loop:

By 1(ii)            (1-0.431) 133.540 x
                    $10^6$ lb/yr

                    78.39 x $10^6$ lb/yr
                    (31.40% of total plant production)


(iii)Extraction ratio (based on nominal throughput):

$$\frac{\text{lb. steam extracted}}{\text{lb. steam turbine input}} \times 100$$

$$\frac{54,120,000}{133.254 \times 10^6} \times 100 =$$

40.61%

(iv) a) Electrical to thermal ratio (based on effective throughput):

$$\frac{\text{kWh output}}{\text{effective steam lb. turbine input}} \times 100$$

$$\frac{6.6625 \times 10^6 \text{ kWh/yr}}{78.39 \times 10^6 \text{ lb/yr}} \times 1{,}000 =$$

84.99 kWh/1,000 lb. steam

b) Based on nominal throughput:

$$\frac{\text{kWh output}}{\text{nominal steam lb. turbine input}} \times 100$$

$$\frac{6.6625 \times 10^6}{133.254 \times 10^6} \times 1{,}000 =$$

50 kWh/1,000 lb. steam

3. Annual Average Operating Parameters

   (i) Nominal Steam Rate (NSR):

$$\frac{\text{nominal turbine steam lb. input} \times 100}{\text{kWh output}} =$$

$$\frac{133.254 \times 10^6}{6.6625 \times 10^6} = 20 \text{ lb/kWh}$$

   (ii) Theoretical Steam Rate (TSR): steam condition 250 psi or 400°F - 1.84"hg:

$$\frac{3.412.75 \text{ BTU/kWh}}{\text{Available energy/lb. steam}}$$

By 3 (vi)

$$\frac{3{,}412.75 \text{ BTU/kWh}}{488.06 \text{ BTU/lb}} = 6.99 \text{ lb/kWh}$$

(iii) Actual Steam Rate (ASR) =

$$\frac{\text{Net Heat Input}}{\text{Gross Heat Input}} \times \text{NST}$$

By 1(i) and 1(ii)

$$\frac{94 \times 10^9}{160.1 \times 10^9} \times 20 \text{ lb/kWh} =$$

11.74 lb/kWh

(iv) Turbine Efficiency =

$$\frac{\text{TSR}}{\text{ASR}} \times 100$$

$$\frac{6.99}{11.74} \times 100 = 61.64\%$$

(v) Boiler System Efficiency $E_B$:

$$\frac{\text{BTU Steam Output}}{\text{BTU Coal Input}} \times 100 =$$

$$\frac{249.688 \times 10^6 \times 1{,}201.1 \times 100}{28.947 \times 10^6 \times 12{,}810}$$

$E_B$ = 80.92%

Taking 1,201.7 BTU/lb. of steam @ 250 psi and 12,810 BTU/lb. coal

(vi) Thermodynamic or Rankine Efficiency $E_R$:

$$\frac{\text{Available Energy}}{\text{Energy supplied}} \times 100$$

Since 59.39% of total throttle steam throughput falls from the condition 250 psig - 400°F to 1.84″ Hg and 40.61% to 6 psig

Available energy =

$(0.5939) (1,201.7-1,157.1) +$
$0.4061 (1,204.6-68) = 488.06$ BTU/lb

$E_R = \dfrac{\text{Available Energy}}{\text{Energy Supplied}}$ x 100

$\dfrac{488.06 \text{ BTU/lb}}{1,201.7 \text{ BTU/lb}}$ x 100 = 40.61%

(vii)  Overall Thermal Cycle Efficiency:

$E_O = (E_B) (E_R) (E_T) =$

$(0.8092) (0.4061) (0.6164) = 0.2030$

Using annually aggregated operational plant data:

$E_O = \dfrac{\text{Electric BTU Output}}{\text{Coal BTU Input}}$ x 100

$\dfrac{3,412.75 \times 6.6625 \times 10^6}{4,536.7 \times 2,000 \times 12,810}$ x 100 =

19.60%

4. Load Cycles

Average Load = $\dfrac{6.662,500 \text{ kWh/yr}}{876 \text{ hr/yr}}$ = 761 kW

(i)  Average demand steam requirements =

$\dfrac{761 \text{ kW}}{50 \text{ kWh/1,000 lb. steam}}$ =

15,220 lb/hr (38.10% of full one boiler capacity)

(ii) Steam requirement to satisfy base load (625 kW) =

$\dfrac{625 \text{ kW}}{50 \text{ kWh/1,000 lb. steam}}$ =

12,500 lb/hr (31.25% of full one boiler capacity)

(iii) Steam requirement for spring peak load (917 kW) =

$$\frac{917 \text{ kW}}{50 \text{ kWh}/1,000 \text{ lb. steam}} =$$

18,340 lb/hr (45.85% of full one boiler capacity)

(iv) Other than spring season annual peak load (1,161 kW) =

$$\frac{1,161 \text{ kW}}{50 \text{ kWh}/1,000 \text{ lb. steam}} =$$

23,220 lb/hr (58.10% of full one boiler capacity)

(v) Average internal demand of power plant (165 kW) =

$$\frac{165 \text{ kW}}{50 \text{ kWh}/1,000 \text{ lb. steam}} =$$

3,330 lb/hr (8.33% of full one boiler capacity)

## 5. Generating Plant Output

(a)     Internal power plant electric consumption.

Assume electric generation required the use of a cooling tower circulating pump (40 Hp), 2 cooling tower fans (40 Hp each) and a condensate return pump (25 Hp). Assume the cooling tower circulating pumps operate 24 hr/day; the cooling tower fans operate 1,000 hr/year; the condensate pump operates 6 hr/day, load factor = .75

Internal consumption =

$$(\frac{40 \text{ Hp}}{.89} \text{ x } 24 \text{ hr/day x } 365 \text{ day/yr} +$$

$$(2 \text{ x } \frac{40 \text{ Hp}}{.89} \text{ x } 1,000 \text{ hr/yr} + \frac{25}{.88} \text{ x}$$

6 hr/day x 365 day/yr))

$$.75 \text{ x } .746 \frac{\text{kW}}{\text{Hp}} = 305.381 \text{ kWh/yr}$$

(b)    Net Electric Output =

(6,662,500 - 305,381) kWh/yr =

6,357,119 kWh/yr

## 6. Power Generation Costs

### (a) Operational costs

(i)  Fuel cost from 1(v) & 5(b):

$$\frac{\$178,746}{6.4 \times 10^6 \text{ kWh}} \times 100 = 2.7929\text{¢/kWh}$$

(ii) Maintenance unit cost (From App. A):

$$\frac{\$3,624}{6.4 \times 10^6 \text{ kWh}} \times 100 = 0.0566\text{¢/kWh}$$

(iii) Part Replacement unit cost (From App. A):

$$\frac{\$2,000}{6.4 \times 10^6 \text{ kWh}} \times 100 = 0.0312\text{¢/kWh}$$

(iv) Makeup water & water treatment unit cost:

$$\frac{\$11,062}{6.4 \times 10^6 \text{ kWh}} \times 100 = 0.1728\text{¢/kWh}$$

Total Operational Cost =

$$\frac{\$195,432}{6.4 \times 10^6 \text{ kWh}} = 3.0536\text{¢/kWh}$$

(b)    Capital cost

Assume straightline depreciation of a $2,500,000 installed cost of the power generation system over its 40-year operational life (and no salvage value at the end) capitalized on the basis of 6-½% internal rate of return:

The annual capitalized cost:

$$A = P \frac{i (1 + i)^n}{(1 + i)^n - 1}$$

Taking $n = 40$, $i = 0.065$, $p = \$2,500,000 =$

$$\frac{(\$2,500,000)\,(0.065)\,(1.065)^{40}}{(1.065)^{40} - 1} =$$

$176,734/yr.

Unit capital cost:     $\dfrac{\$176,734 \times 100}{6.4 \times 10^6 \text{ kWh}} =$

2.7614¢/kWh

(c)   Generating cost

Including capital cost:   (a) + (b) =

$372,166/yr. or 5.8151¢/kWh

7. Purchase Cost of Electricity

We use Illinois Power Company's local rate structure #21 and assume that switchover from absorption chiller system to electrical one would place an additional power demand of 60 kW during cooling season.

(a)   Demand Charge

| Month | Facility Demand (kW) | | Rate $/kW | Demand Charges ($) |
|---|---|---|---|---|
| June | | 950 | x  10.47 | 9,947 |
| July | *Summer | 1,250 | x  10.47 | 13,088 |
| Aug. | *Summer | 1,150 | x  10.47 | 12,241 |
| Sept. | | 1.100 | x  10.47 | 11,517 |
| Oct. | | 1,125 | x  4.97 | 5,591 |
| Nov. | | 1,100 | x  4.97 | 5.467 |
| Dec. | | 1,100 | x  4.97 | 4,970 |
| Jan. | | 1,050 | x  4.97 | 5,219 |
| Feb. | | 1,125 | x  4.97 | 5,591 |
| March | | 1,125 | x  4.97 | 5,591 |
| April | | 850 | x  4.97 | 4,225 |
| May | | 950 | x  4.97 | 4,722 |
| Total Demand Charge | | | | $88,169 |

*By definition of the utility contract.

(b)    Energy charge
- For the first 100,000 kWh/mo. @ $0.0414/kWh:
  12 x 100,000 x $0.0414 = $49,680
- For the rest @ $0.033/kWh:
  $(6.4 \times 10^6 - 1.2 \times 10^6)\,(\$0.033) = \underline{\$171,600}$

    Total                          = $221,280

(c)    Transformer charge
    (2,000) kW x $0.48/kW x 12 = $11,520

(d)    Off-peak consumption credit @ 1¢/kWh, taking ratio of off-peak time to on-peak time as 2.0:
    $(\$0.01/\text{kWh})\,(2/3)\,(6.4 \times 10^6 \text{ kWh}) = \$42,667$

(e)    Total purchase cost including tax @ 5.08%:
    (1.0508) (a + b + c - d) = $292,440

(f)    Unit cost of electricity if purchased:
$$\frac{\$292,440}{6.4 \times 10^6 \text{ kWh}} \times 100\text{¢}/\$ =$$
4.5694¢/kWh

## 8. Cost Comparison

(a)    Cost savings when generating cost *excludes* capital cost:
    From 6(e) & 7(e)

        $292,440 - $195,432 = $97,008

    % Savings:    $\dfrac{\$97,000}{\$292,440} = 33.17\%$

(b)    Cost savings when generating cost *includes* capital cost:
    From 6(b) & 7(c)

        $292,440 - $372,166 = $79,726

## 9. Costs With Natural Gas

Used as alternative fuel: If natural gas were used at its current cost of \$3.846/million BTU (MBtu) assuming that all other conditions remain the same the fuel cost contribution to unit cost of kilowatt hour of electricity would be:

$$\text{Fuel Cost} = \frac{\$3.8460/\text{MBTU nat. gas}}{\$1.5379/\text{MBTU coal}} \times \frac{2.7929}{\text{kWh}}$$

6.9845¢/kWh

Operational cost = (6.9845 + 0.0566 + 0.0312 + 0.1728)¢/kWh = 7.2452 ¢/kWh

Percentage excess cost over purchase price:

$$\frac{7.2452 - 4.5694}{4.5694} \times 100 = 58.6\%$$

## REFERENCES

1 Bernhardt, G.A. Skortzki and William A. Vopat, *Steam and Gas Turbines*, McGraw-Hill Book Company, Inc., First Edition, 1950.

2 Bernhardt, G.A. Skortzki and William A. Vopat, *Power Station Engineering and Economy*, McGraw-Hill Book Company, Inc., 1960.

3 J.K. Salisbury, *Steam Turbines and Their Cycles*, John Wiley & Sons, Inc.

4 Joel D. Justin and William G. Mervine, *Power Supply Economics*, John Wiley & Sons, Inc.

5 Roy Meoder, *Cogeneration and District Heating*, Ann Arbor Science Publishers, Inc., 1981.

6 U.S. Department of Energy, Division of Fossil Fuel Utilization, "Cogeneration: Technical Concepts-Trends-Prospects," Report DOE-FFU-1703.

# CHAPTER 19
# Cogeneration Planning at American-Standard

*M.A. Mozzo, Jr.*

The term "cogeneration" is rapidly becoming a new buzzword in the energy manager's vocabulary. Many managers are actively pursuing cogeneration as part of their energy programs. This chapter presents American-Standard's views of the potential for cogeneration at its plants, and considerations that must be evaluated in such projects.

## WHY COGENERATE?

There are four key factors which are renewing the interest in cogeneration systems. First and probably foremost is rising electrical costs. In the 1970's and into the 80's, rising fuel, materials, and labor costs have dramatically increased electrical costs. Further affecting electric costs has been the high cost of bringing new plants on line. Inflation has affected construction and financial carrying costs for those utilities engaged in a construction program.

In some instances, completion of a new plant and inclusion of the entire cost of the plant in rate base at one time could result in rate increases of at least 30 percent and probably much more. Because of the steepness of the rate increases, public utility commissions are reluctant to pass them on to favored residential customers. Consequently, much work is being done on cost-of-service and rate designs to the detriment of high-load industrial customers. Such discriminatory rate efforts, though, make cogeneration for industry even more economically feasible.

A second factor which is sparking interest is the Public Utilities Regulatory Policies Act of 1978 (PURPA). Essentially PURPA was intended to encourage industrial use of cogeneration through rate incentives. A group of utilities, however, challenged PURPA and the subsequent implementing rules of FERC regarding avoided cost payments to qualifying cogenerators and interconnection requirements. After a series of court battles, the U.S. Supreme Court upheld FERC's rulings in 1982. While the Supreme Court's decision has alleviated uncertainty about cogeneration at the Federal level, the battle over avoided costs has shifted to the state level. Decisions are now being made at this level between cogenerators, utilities, public utility commissions, and even the courts as to the proper contract rates and requirements for qualified cogenerators.[1]

The third factor affecting cogeneration activity is technology. Recent developments have been made to both increase efficiencies of equipment as well as to provide smaller equipment which is sized closer to most industry needs and at an economical price. These actions will increase market penetration.

In a recent *New York Times* article (June 10, 1984),[2] it was estimated that cogeneration will increase from 5% of total U.S. electricity generated in 1983 to 15% by 2000, a three-fold increase. Sales for cogeneration equipment will increase to almost $5 billion/year in 2000 as contrasted to under $½ billion in 1980.[3] Technological developments will indeed make cogeneration available to more consumers.

The fourth factor which will have an impact on cogeneration is the future of the U.S. electric utility industry. Frequently in the news today are stories of utilities in financial difficulties because of the costs of building large central power plants. Clouding the financial issues are the uncertainties of acid rain legislation and the future of nuclear power plants. There is a reluctance on the part of utilities to begin new plants.

Utilities, aided by public utility commissions, espouse the benefits of conservation with the goal of reducing the need for capacity additions. Alternate plants powered by solar, wind, and low-head hydro are also being built, though on a very small scale, as an adjunct to conservation.

Cogeneration is another alternate technology mentioned to reduce capacity addition requirements. This action by utilities and public utility commissions may then spur cogeneration technology; however, we believe it will be more a development of necessity than one of an alternate technology. While capacity reserves remain high today, demand is still growing at 2-3% per year. Spot shortages and rising electrical costs may well spur small, industrial cogeneration sites more for survival than by economics.

## COGENERATION TECHNOLOGY

Basically, there are three types of cogeneration systems. These are:

1. Steam Turbines: High-pressure steam is generated in a boiler and then is piped to a non-condensing steam turbine to generate electricity. The extracted steam is used in the plant for process and or heating at a suitable pressure. Fuel for the boilers can be biomass, coal, fuel oil or natural gas. Capital costs for this system can be high, but operating costs, if biomass or coal is used, can be low.

2. Diesel Engine Generator: Uses a reciprocating internal combustion engine burning fuel oil or natural gas. Advantages to this system are low capital requirements and high cogeneration efficiencies. Reliability, maintainability and operating costs are a negative. Cost and supplies of natural gas and fuel oils are key negative factors in using diesel generators.

3. Gas Turbine Generator: Uses a combustion turbine/generator burning a light fuel oil or natural gas. This system has a low capital requirement as well as excellent reliability and maintainability. Operating costs however can be high because of fuel requirements and efficiencies.

At the 6th AEE Energy Audit and Management Symposium, Mr. Macauley Whiting, Jr. of Decker International Inc. presented a paper on small cogeneration systems. In this paper, Mr. Whiting presented a table of factors which affect the economic attractiveness of cogeneration. He concluded that the two most important factors in evaluating cogeneration feasibility were 1, the cost of the electric being replaced by the cogeneration system, i.e., it should be high, and 2, the cost of the fuel used in the cogeneration system, i.e., it should be low.[4]

We at American-Standard fully agree with this conclusion, though we feel that proper balance of electric and thermal loads is equally important. Of these three factors, we feel that for our plants the most controllable one is the cost of the fuel to be used in the cogeneration system.

We have, currently, three plants served with gas from American-Standard solely owned wells through our privately owned gathering system, i.e., pipeline. Our conclusion then is that the best cogeneration system for our needs is a gas turbine fired by a natural gas supply which we control both the quantity and price. To this end then, we have, at this point of time, essentially limited any in-depth study of cogeneration to the three major facilities where we have gas wells and the private pipeline system available.

## A CASE STUDY – BACKGROUND DATA

This section reviews the economic feasibility of cogeneration at one of the three facilities mentioned previously. Key to this study is the existence of sufficient private natural gas supplies, plus the private pipeline system to support the gas turbine(s). The pipeline in this case was built to provide gas supplies for other process needs, but it was sized to provide expansion requirements such as cogeneration.

The facility uses electricity year-round for process needs typical of a manufacturing facility, i.e., lighting, motors, ventilation equipment, etc. It also uses natural gas in boilers for both process and heating requirements. Boiler use is year round because of the process requirements. Table 19-1 shows the billing determinants for electric and natural gas for an historical year. Costs during this same period were $662,213 (electric) and $626,817 (boilers) or a total of $1,289,030.

For purposes of conducting an initial economic feasibility, data were extracted from the *AGA Manual, Cogeneration Feasibility Analysis*. For purposes of this study, an 800-kW gas turbine unit was selected. Basic data for this unit:

Capital Costs (Dec 1983) – $1040/kW (Grid Connect)
                          – $ 940/kW (Isolated)
Fuel Rate – 0.168 Therms/kW (Full Load)
          – 0.194 Therms/kW (Part Load)
Heat Available – 0.1008 Therms/kW (Full Load)
              – 0.1164 Therms/kW (Part Load)

### TABLE 19-1. Base Plant Billing Determinants

| Month | Demand (kW) | Electric Energy (kWh) | Natural Gas Boilers (MCF) |
|-------|-------------|-----------------------|---------------------------|
| Jan | 2,400 | 1,449,600 | 12,317 |
| Feb | 2,400 | 1,396,800 | 11,174 |
| Mar | 2,460 | 1,483,200 | 13,220 |
| Apr | 1,520 | 1,413,600 | 10,120 |
| May | 2,400 | 1,425,600 | 8,831 |
| Jun | 2,520 | 1,485,600 | 8,650 |
| Jul | 2,580 | 1,308,000 | 5,273 |
| Aug | 2,580 | 1,204,800 | 5,355 |
| Sep | 2,520 | 1,502,400 | 8,999 |
| Oct | 2,580 | 1,464,000 | 8,325 |
| Nov | 2,520 | 1,555,200 | 10,408 |
| Dec | 2,460 | 1,404,000 | 12,972 |
| TOTALS | — — | 17,092,800 | 115,194 |

Several assumptions had to be made in order to complete the calculations for this study:

1. Energy usage patterns would remain the same.
2. Cost/savings data generated for historical year (1983) used to calculate simple payback only.
3. Supplemental gas requirements for boilers to be purchased from private pipeline at cost of $5.00/MCF.
4. Incremental gas cost for cogeneration application is $2.00/MCF.
5. Electric utility is willing to provide backup service, if needed, at a cost of 70% of demand requirements.
6. Operating and maintenance costs for the turbine generator at 0.75 cents/kWh.
7. Boiler efficiency is at 80% average for the time period.

## CASE STUDY – BASE LOADED ELECTRICALLY

In this situation, one 800-kW generator is installed and run constantly (8,760 hours per year). Electrical requirements reduce the monthly demand to a range of 1600 to 1780 kW and total yearly energy consumption to 10,084,800 kWh. Total electrical costs will then be $417,297 using the 1983 rate structure, fuel adjustments, and taxes.

Other calculations show a capital investment of $832 k, gas turbine fuel input at 115,426 MCF of gas, available waste heat recoverable at 69,256 MCF equivalent, and supplemental heat required at 38,014 MCF of natural gas.

Cost Summary:

| | |
|---|---|
| Investment Cost: | $832,000 |
| Operating Costs: | $942,252 |
|     Purchased Electric | $417,297 |
|     Standby Demand | $ 51,475 |
|     O & M | $ 52,560 |
|     Boiler Fuel | $190,070 |
|     Gas Turbine Fuel | $230,852 |
| Previous Utilities | $1,289,030 |
| Net Savings | $346,776 |

In this case the simple payback savings using current investment and operating costs and savings will be 2.4 years ($832,000/346,776).

## CASE STUDY – TOTAL INDEPENDENCE

In this case, four 800-kW generators are installed to provide necessary power and backup for electrical usage, i.e., no electric utility interconnect. Some supplemental heat will be required for boiler needs. Based on electric load requirements, it is assumed that two generators will be run at full load and another unit at part load. Based on this operating scenario, calculations show a capital investment of $3,008 k, gas turbine fuel input at 289,333 MCF of gas, available waste heat recoverable at 173,600 MCF equivalent, and supplemental heat required of 8,640 MCF of natural gas.

Cost Summary:

| | |
|---|---|
| Investment Cost: | $3,008,000 |
| Operating Costs: | $ 750,062 |
| O & M | $ 128,196 |
| Boiler Fuel | $ 43,200 |
| Gas Turbine Fuel | $ 578,666 |
| Previous Utilities | $1,289,030 |
| Net Savings: | $ 538,968 |

Simple payback for the total independence case is 5.6 years ($3,008,000/538,968).

The lack of balance between electrical loads and boiler or thermal requirements adversely affects the savings for the total independence case. Certain factors, which if changed, would affect the outcome of this case. For example, if electric costs were to go up 20%, from about 3.88 cents/kWh to 4.65 cents/kWh, simple payback would be reduced to 4.5 years. Simple payback could also be reduced if the incremental gas cost for turbine fuel is lowered. For example, if the cost was $1.00/MCF instead of the $2.00/MCF assumed, then the simple payback would be 3.6 years.

## SUMMARY

Calculations made in the case study, that of a small installation (less than 10 MW), show the effects of three key factors or elements in a cogeneration project: 1) the balance between electrical and thermal loads, 2) the price of utility purchased electricity, and 3) the price of the cogeneration system fuel source. All three items are important for a cogeneration project to be economically feasible.

## REFERENCES

1 "Cogeneration Potential Projected to Double in Capacity by 2000, Energy Analysts Project," *Energy Users Report*, p. 474, May 31, 1984.

2 "Power Industry's Uncertainty," *New York Times*, May 22, 1984.

3 "Cogeneration Jars the Power Industry," *New York Times*, June 10, 1984.

4 "Guidelines For Assessing The Feasibility Of Small Cogeneration Systems," a paper presented by Macauley Whiting, Jr., Vice President Decker Energy International Inc. at the 6th AEE Energy Audit and Management Symposium, June 1984.

5 *AGA Manual Cogeneration Feasibility Analysis,* April 1982.

# INDEX